Aerobic Wastewater Treatment Processes

Aerobic Wastewater Treatment Processes
History and Development

David F Tilley

Publishing
London • New York

Published by **IWA Publishing**
Alliance House
12 Caxton Street
London SW1H 0QS, UK
Telephone: +44 (0)20 7654 5500
Fax: +44 (0)20 7654 5555
Email: publications@iwap.co.uk
Web: www.iwapublishing.com

First published 2011

British Library Cataloguing in Publication Data
A CIP catalogue record for this book is available from the British Library

Library of Congress Cataloging-in-Publication Data
A catalog record for this book is available from the Library of Congress

ISBN: 9781843395423
ISBN: 1843395428

Contents

About the author

The author, David Tilley, has had many years of experience in sewage treatment, trade effluents and the water industry generally.

He began his working life in 1962 in the City Analyst Laboratories, Coventry Corporation, and was later appointed Assistant Chemist, Sewage and Trade Wastes.

He then gained a post in the field of trade effluent control, first in Coventry and then for Avon Division of Severn Trent Water Authority. He has been employed in various capacities within the Severn Trent Water Authority, later Severn Trent Water plc, until 1992.

In 1993 he gained the post of Principal Consultant with Environmental Management Associates Ltd (EMA Ltd), an independent consultancy specialising in sewage treatment, industrial wastewater treatment and water related issues. EMA Ltd have undertaken a wide range of assignments for both private and public sector interests in the UK, Europe and North Africa.

David Tilley is a qualified professional and holds a post-graduate research degree of Master of Philosophy and a Diploma in Water and Environmental Management. He is also a Fellow of the Chartered Institute of Water and Environmental Management, a Fellow of the Royal Society of Health and a member of the European Water Pollution Control Association.

Acknowledgements

Firstly I must thank my wife, Herta, for her continued support and help in editing this book.

Invaluable advice and assistance was provided the late Dr J. A. Anderson, former Principal Lecturer in Civil Engineering, Coventry University.

I am indebted also to the late Mr C. T. Gilbody who gave me early inspiration and direction.

My thanks also go to Dr P. Fox, Dr M. L. Hyde, Dr M. E. Brown, Mr P. Starkey, Mr P. Fell, Mr A. Q. Khan OBE, Mr R. Hunt, Mr S. Weston, Mr A. Pohl, Miss E. Marriott, Mr J. Tilley, Mrs M. Haynes and many other valued friends for their support and encouragement.

In addition I would like to express my thanks to Maggie Smith (Editor) and Michelle Jones (Production Manager) of IWAP Publishing for their invaluable help in preparing this book for publication.

Chapter 1
Introduction

BACKGROUND

The aim of this book is to examine the implications and development of what are, essentially, the two main processes of *aerobic* sewage treatment, namely the bacteria bed or biological filter and the activated sludge process. These principal forms of rendering sewage more amenable to disposal have been extended and modified over a considerable time period. *Anaerobic* treatment is not covered as it has only been used widely in most recent times, mainly for sludge treatment. The most common treatment by septic tank employs anaerobic principles but only for the first stage of treatment. Treatment is completed by organisms and bacteria in the sub-soil of the soak-away.

Anaerobic treatment was developed primarily for disposal of waste sludges and bio-solids, which are by-products of aerobic sewage treatment. It is a process used mainly to reduce the volume of waste solids for ultimate land-fill, spreading on farm land or composting. Anaerobic digestion was found to be a useful source of methane gas (this has a similar calorific value to natural gas), which on larger sewage treatment works is used to generate electricity. For this reason anaerobic digestion is often used as a gas-generating process and any reduction in solid matter is incidental. It is hardly ever used as a stand-alone treatment process except for certain concentrated food wastes or farm slurries. All municipal sewage works utilise aerobic treatments as do all smaller package plants. As stated, larger plants may use anaerobic digestion but only to reduce bio-solids or to generate methane gas. Since it is not a widely adopted treatment method and is used principally to reduce the solid by-products of aerobic treatment, it is therefore not included as a topic in this book.

Other factors, such as advances in microbiology and analytical chemistry, the need for potable water, liquid industrial waste treatment and pollution legislation have all influenced and encouraged further improvement to these systems which are now very sophisticated.

From the time man settled in tribal communities, human waste has presented a disposal problem. Nomadic communities were able to leave their waste behind and proceed to new areas. For any civilisation to develop and thrive, there is a need for stability, and with few exceptions, nomadic societies have had little influence on man's cultural or social development.

In early civilisation the accumulation of human and animal waste where small communities became established did not usually cause serious health problems. The volume of waste produced, and hence the resulting impact upon health, would tend to be related to intensity of population and later to water supply and hence waste water. The latter was particularly so for water-carriage waste systems which were developed in Europe and North America.

Historically there has been little interest in the subject of sewage which perhaps is not surprising as it was often regarded as a 'taboo' topic. This subject has been neglected by the historian and archaeologist too, such that even today, historical knowledge of sewage treatment disposal is largely anecdotal. Much of the available historical records can often be traced back to the same few sources.

The existence of sewers and the need to dispose of, or treat, sewage would only be necessary in stable, civilised societies where urban population reached certain levels. This would be a further explanation for the lack of historical information on this subject. Only in our comparatively recent history have urban populations assumed such proportions. During the reign of Queen Elizabeth I, London doubled its population to 200 000. The population of the whole country at that time was some four million.

It is most important to recognise the difference between *disposal* which usually infers the transportation and conveyance or removal of waste, and *treatment* which may be physical, chemical or biological and degrades or changes the nature of the sewage to be more acceptable and less noxious or polluting. Many historical records describe the existence of drains and sewers and water supplies, but make no mention of 'treatment'. The very existence of such collection systems is often mistakenly used to infer treatment. There are, fortunately, some exceptions.

EARLY HISTORY

The earliest reference to drainage must be the sewers of the Akkadian 'cities of brick', built in the valley of the River Tigris close to what is now Baghdad, around 3000BC (Wylie, 1959; p. 21). These dwellings reportedly all had bathrooms and drains which carried sewage to soak pits. Interestingly, these soak pits were emptied only after 'the very substance of the earth which the sewage had contained, had perished' (McCallum,1968; p. 8). This implies a crude composting treatment process and is surely the earliest record of any form of centralised sewage treatment.

Later, around 1600BC, on Crete, the Minoan civilisation was reaching its peak. (McCallum,1968; p. 10). The Royal Palace in the Minoan capital of Knossos enjoyed a pressurised, piped, water supply and a cement-jointed, stone sewerage system (Sprague de Camp, 1963).

The Minoan system was very innovative and depended on a 'constant-flow' system for flushing. The lavatory was built over a channel through which water ran constantly. These latrine drains were generally kept separate from other drains to prevent foul gases rising through all the system's inlets (Reynolds, 1943). Dry refuse was deposited in large pits called 'Kouloura' and earth was added to form a layered system. Water was sprayed on this and the waste matter composted before being applied to local farmland. This could be regarded as an example of conscious waste *treatment* (and not *disposal*) in which the nature of the waste is changed and indeed, utilised, not merely transported away.

There is reference to sewage disposal in the Bible by Moses, who formed a most comprehensive code of health for the ancient Israelites. The quotation from Deuteronomy Chapter XXIII, verse 12: 'When thou wilt ease thyself abroad, thou shalt dig therewith, and shalt turn back and cover that which cometh from thee'. This practice is now known to have a sound scientific basis and relies on soil bacteria to reduce the organic matter to harmless (and useful) end-products.

THE ROMANS

No historical record of sewerage can be complete without mention of the Roman influence on such matters. Roman engineering, like that of the earlier empires, was based upon the intensive application of basic principles with an abundance of raw materials, cheap labour, and projects carried out over a long time period. Rome had plentiful supplies of brick, stone, and timber. Labour was especially cheap during the last century of the Republic because the expansion of the Roman Empire had supplied Rome with thousands of slaves.

The Romans devoted much of their energy and capital to useful public works. A governor or senator may have been disreputable in many respects, but he thought that honour could be gained by presenting the people with some useful public work. Under the Republic this was a method of bidding for votes and even after elections had ceased the tradition lived on. It was an underlying principal of government and is still manifest today.

Although the Romans inherited some of their engineering ideas from the Etruscans, they adopted many more from the Greeks and Orientals who came to Rome, voluntarily or as captives. It has been said that the Romans contributed practically nothing to pure science. Abstract studies in philosophy or mathematics were developed by the Greeks. Rome produced almost no research scientists, but many remarkable generals, statesmen, and administrators. In the applied sciences, however, the Romans excelled. Roman architects and engineers made far-reaching advances towards improving the conditions of everyday life. These improvements were established in other countries as the Roman influence spread with the development of the Roman Empire.

The sewers of Rome were begun in the days of the Tarquins (616 to 510BC), when a ditch was dug to drain the swampy land between the seven hills. The first sewer followed the course of an existing stream from the Forum of Nerva to the River Tiber, and was a stone-lined channel to carry off storm waters. Successive generations enlarged and improved this drain, covered it with a stone barrel vault, and led effluents into it. It became the Cloaca Maxima, so large that it was possible to row through it in a boat.

The Cloaca Maxima remained primarily a storm water sewer system with numerous openings in the streets to drain off rain water. By the time of the Principate (509BC), a number of large public latrines had been built and connected to this sewer system. Used water from public baths and industrial establishments, such as dyers and tanners, was channelled into these appliances and provided a 'flushing' effect by maintaining a flow even in dry weather.

Although this latrine system was a great advance from the conditions of primitive times, it was still a long way from modern sanitation. Many dwellings of this period had their own latrines, nearly always on the ground floor and often connected with the sewer system. A few (at least at Ostia, Herculaneum, and Pompeii) had sewer connections to their upper storeys. The majority of Romans, however, lived either at a distance from the latrines or on the upper floors not furnished with lavatories. Those without this facility either carried their sewage to cesspools or, although there were laws against this, threw it out of the window. So Rome, despite its splendid sewers, was still pervaded by the stench of human excreta.

The Roman sanitary system was regarded as so advanced, however, that in 1842, a British Royal Commission, appointed to consider ways of improving the health of the people in London, included in its report, a description of the sanitary arrangements of the Coliseum and of the Roman amphitheatre at Verona. They were thought, apparently, to be better than anything in Britain at that time.

The sewers of the Roman era are quite famous, but their initial purpose was for the removal of rain-water and ground-water and not as foul sewers (Quennell and Quennell, 1977; p. 96). The filth of Rome was consequently discharged to the River Tiber. Although the Romans propagated the use of baths and drainage systems throughout their empire, with the decline of Rome, sanitation became neglected. This is almost certainly the reason for the increased incidence of disease at this time and the marked depopulation of Rome and other Roman cities which resulted (Buer, 1926; p. 104).

THE MIDDLE AGES

There was no improvement in methods of waste removal in Europe for many centuries. With the advent of the Dark Ages little progress was made for 1400 years, from the early fifth century until the beginning of the nineteenth century (Buer, 1926; p. 104).

In the Middle Ages all refuse including human excreta was thrown into the street. This practice was often forbidden and bye-laws were passed to prevent such action, but little progress was made. The practice of throwing waste from windows and houses into the streets was forbidden in fourteenth-century Paris. The Coutume de Paris forbade this practice in 1372 and again in 1395. (Quennell and Quennell, 1977; p. 101). It was only in 1513 that every house was ordered to have a privy or cess pit or area where such waste could be deposited (Quennell and Quennell, 1977; p. 101). This was well meant but proved impossible to enforce.

In the eighteenth century most continental towns still threw human excreta and garbage into the street. In Madrid even the Royal Residence had no privies in 1773. In England both Richard II in 1388 and later Henry VIII in 1531, attempted to prevent water pollution due to cesspool discharges (Buer, 1926; p. 99). Medieval Paris had open ditches intended for use as storm drainage and although it was disallowed, Parisians dumped household waste into these drainage systems. There was seldom sufficient rainfall to completely flush out such ditches and they rapidly became very dirty and smelly.

Around 1400 Hughes Abriot, who was a provost of Paris in the time of Charles VI, roofed over certain stretches of these ditches to form sewers. After his death the remaining open ditches were gradually covered (McCallum, 1968: p. 10) but Paris still lacked a proper sewerage system. In 1550 Henry II repeatedly tried to get the Parisian parliament to build new sewers but he was not in favour of paying for such work from the Royal Exchequer. Despite the increase in stench and disease, this situation prevailed until, in the seventeenth century, Louis XIII began a programme of sewer building (Quennell and Quennell, 1977; pp. 102–103).

THE EIGHTEENTH CENTURY

The extensive works of modern Parisian sewerage is only about 160 years old, having been built mainly under the auspices of Napoleon III. In 1752 in London it was recorded that privies were well regulated (Buer, 1926; p. 101). This referred to the practice of emptying and removing piles of sewage and domestic waste which had accumulated in peoples' courtyards or outside their houses. The rich had private garden areas where they deposited their waste. Such waste was removed regularly by people who were termed 'night soil men' or 'scavengers'.

Around any large expanding town there was a great need for a ready food supply. The nearby cultivation of vegetables and fruit for food resulted in eighteenth-century London being surrounded by market gardens. These gardens required large quantities of manure and at this time it was no longer necessary to pay scavengers and night soil men to remove refuse matter. They were eager to collect such material for use in the market garden industry. This practice of using sewage as fertiliser was developed later in a different form (see Chapter 4).

The scavenging of refuse from the streets and private areas of larger houses was carried on throughout medieval times and, as stated, prevailed until the mid-1700s. If properly carried out, the methods of the mid-eighteenth century were not necessarily highly insanitary (Buer, 1926; p. 101). It is thought that scavenging and regular removal of accumulated refuse matter, including human waste, for subsequent disposal on to agricultural land to aid crop growth was not altogether a bad thing.

Early drainage schemes were devised, as in Roman times, to prevent flooding. London and Bristol were quick to construct elaborate schemes for dealing with flood water. These schemes involved the building of surface water drainage systems which conveyed water from the built-up areas to the nearby rivers. Often such drainage consisted mainly of gullies down the centre of streets and flow was aided by the collection of roof water and discharge through gutters and down-pipes which was a technique which gradually became established in built-up areas. As the cobbled streets were seldom kept in good repair these measures were often not very effective. With the introduction of surface drainage systems, which was the original intended use of these urban sewerage schemes, drains became a ready and obvious receptacle for waste matter from nearby houses. All kinds of refuse was thrown into these drains and so blockages became a common occurrence.

The association of waste disposal with the advent and provision of sewerage, albeit surface water sewerage, within towns, led to the next development which was closely associated with the provision of water supply and the invention, and popular use, of a water carriage system for sewage disposal.

To discuss the further progress of sewage disposal, and eventual treatment, it is necessary to consider two of the most important developments which occurred at around the same time. These far-reaching changes to social behaviour were the supply of piped water and the invention and propagation of the water closet as a means of disposing of faecal matter by a water carriage system. It may be argued that the removal and disposal of solid waste as manure, was probably of more benefit than the catastrophic results of disposal of sewage by a water carriage system into nearby water courses. The water carriage system had the effect of polluting drinking water sources and inevitably led to widespread outbreaks of plague and disease.

To a certain extent the one relied heavily upon the other for its mode of operation and the eventual fate of water-borne waste had an effect on the quality of the piped water supply. These two developments are inter-related and eventually led to the treatment of sewage as we know it today.

Chapter 2

Water supply

WATER SUPPLY

The water supply in London depended originally on small tributaries of the Thames such as the Fleet and the Walbrooke and several wells. The ordinary houses did not have a piped water supply and water was normally transported by servants from nearby wells or public water supply pumps or public fountains. The poor would have had to fetch their own from these sources. Some pumps still remain in inner London, but are no longer in use (there is one in Bedford Row).

As urban conurbations grew larger and populations increased, these sources of supply began to fail, and more water was brought into the city via lead pipes from tributaries of the river and from the River Thames itself. A patent for a leaden conduit for conveying water from the Tyburn Brook to the City of London was granted as early as 1236 (Stanbridge, 1976; vol. 2, p. 9).

Even at the beginning of the seventeenth century the citizens of London were almost entirely dependent for drinking water on what was carried to them from the Thames. This was done by the Honourable Company of Water Tankard Bearers, who had the right to sell the water under Charter. (Stanbridge, 1976; vol. 2, p. 9). These were tradesmen who carried water around the streets of London in large barrels, or other containers, on horse-drawn carriages or carts and often on yokes across their shoulders. The poor could not afford this service and as a result often went unclean and drank little water.

At a later date other conduits were laid from Dalston and Bloomsbury to supply public fountains where water supplies could be obtained by local inhabitants. These were gravity-flow systems and involved no pumping of water.

The richer citizens would convey water by pipes from the public conduits to their houses. This inevitably led to a reduced public supply and often led to protests. Another source of complaint was that thoughtless persons found the sources of the water conduits a convenient dumping place for rubbish, sometimes even to the point of blocking them entirely.

LONDON BRIDGE WATER WORKS

By the end of the sixteenth century provision of an adequate water supply to London was becoming a matter of some urgency. The situation was helped to some extent by a scheme to pump water from the River Thames. A water wheel fixed to one of the arches of London Bridge was used to drive a pump to supply water to the surrounding areas. This was the invention of Peter Moriris (Quennell and Quennell, 1977; pp. 103–105) a Dutchman, who had considerable experience of pumping water (Figure 2.1).

Figure 2.1 Water works built under London Bridge by Peter Morris in 1582 (Copyright 2011 Mills Archive Trust, reproduced with permission).

He was granted a lease of one arch of the bridge for a sum of ten shillings a year for a period of 500 years. London's Court of Common Council granted this lease to Morice in December 1582, so the contract will last until the year 2082. He later leased further arches for more water-wheel-powered pumps, and formed the London Bridge Water Works.

His water wheel, revolved at a rate of 6 revolutions per minute, operated two force pumps which lifted river water to a storage cistern on top of a 128 foot tower (Figure 2.2). The water passed through a coarse screen and then flowed by gravity through wooden mains serving several streets. Lead pipes were tapped off the wooden main to serve individual dwellings. This is thought to be the earliest example of a pumped water supply serving domestic dwellings.

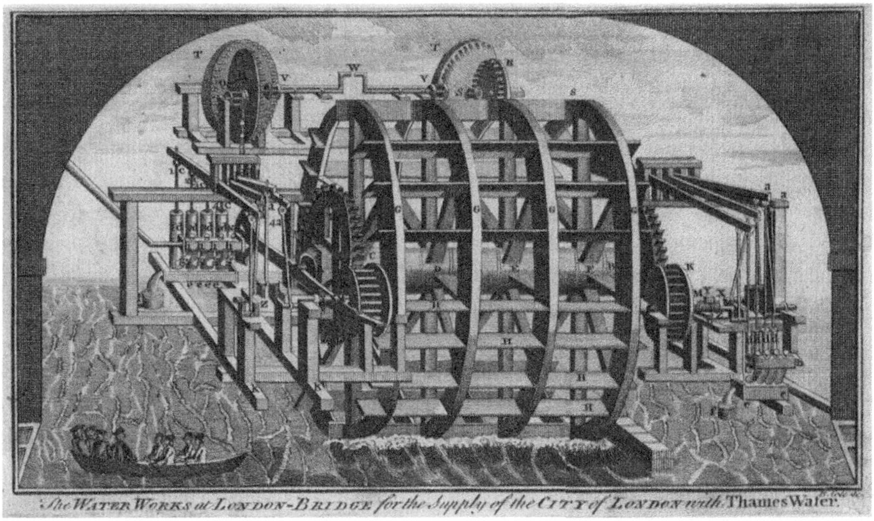

Figure 2.2 London Bridge water wheel mechanism 1635 (Copyright 2011 Mills Archive Trust, reproduced with permission).

By 1822 the London Bridge Water Works was supplying 727 m^3 (160,000 gallons) of water per hour. The water wheels and pumps were, by this time, blocking five arches of the bridge. These structures were now becoming a serious navigation hazard to craft passing under London Bridge. The bridge itself was becoming unsafe and was to be replaced. Consequently, and with some pressure from the rival New River Company, the London Bridge Water Works was removed by an Act of Parliament of 1822. The shares of the London Bridge Water Works were transferred to the New River Company.

THE NEW RIVER COMPANY

It was from a public health point of view that seventeenth century London, like ancient Rome, was compelled to seek additional water supplies from a distant and higher source. Local wells were mostly polluted, the Thames water was also polluted, and no pumps then invented could supply London adequately with water.

The New River Company was a scheme originally devised by a Captain Edward Colthurst of Bath, who planned to build a 'new river' from Chadwell and Amwell in

Hertfordshire to a storage reservoir in Islington, London (Quennell and Quennell, 1977; pp. 103–105). He ran into financial difficulties with the project but was supported by Hugh Middleton who was a wealthy member of parliament, an entrepreneur and former goldsmith and textile merchant. Middleton provided capital, took over the scheme and appointed Colthurst as a partner and overseer. The venture met much resistance and again ran short of money until King James I put up half the outstanding debts in return for a half share in the profits. Middleton, with the help of this royal support eventually succeeded in completing the scheme in 1613.

The scheme was a great feat of civil engineering for its time. The water was conveyed by constructing an artificial river, mostly in open channel which generally crossed valleys at their head, although some aqueducts were also used. There was a gradual fall of 18 feet over a length of 38 miles. The channel discharged water into Round Pond near Sadlers Wells, later called the New River Head. It was 80 feet above the level of the Thames at this point which was ideal for distribution by gravity to the surrounding streets.

The Water Tankard Bearers thought that the New River would be detrimental to their trade but this was not so initially. The six-foot-long, elm wood pipes, through which the water was distributed were costly to lay over private land and to have a property connected to the main water system was expensive. Instead, the locals paid the Tankard Bearers (one penny a pail) to bring them water from the New River Head and so their trade thrived.

The springs of Chadwell still form part of the water supply of London but this New River is now carried in an iron pipe and its length has consequently been shortened from over 38 miles to 27 miles.

Even after its construction and commissioning a very large proportion of the city was still supplied with Thames water. During the eighteenth century London continued to grow, not only in terms of population, but also in area. This resulted in large portions of the community being outside any reliable water supply network. The New River Company could not serve many of the new districts and so numerous new undertakings were formed for this purpose. By 1886 nine water companies supplied London with 150 million gallons of water a day, 70 million gallons of this water came from the River Thames and 60 million gallons from the New River Company.

The water supply of London was a matter of private enterprise right up until the twentieth century. Supplying water, however, was not always the profitable undertaking that its promoters had hoped. The shortcomings of many of the water companies were partly due to financial difficulties. Other factors were technical problems which were only slowly overcome, foolish attempts at competition and rivalries between different companies. Such destructive rivalries were not uncommon in civil engineering ventures. Railway companies and canal construction were subjected to similar obstacles.

PUMPING DISTRIBUTION AND STORAGE

There were nine companies founded to deal with the water supply to Greater London between 1669 and 1806 (Stanbridge, 1976; vol. 1, p. 10). Of these, all but two drew their supply from the River Thames and this was made possible only by improved pumps which, by the year 1800, would be entirely powered by steam.

To supply water from a lower to a higher level necessitated the use of pipes instead of open channels. Originally water mains were made of elm trunks (Figure 2.3) with a 150 mm (6 in) diameter hole bored through their length (Figure 2.4). They were fitted end to end and jointed with clay. These mains were surprisingly long lasting but were gradually replaced by more reliable materials.

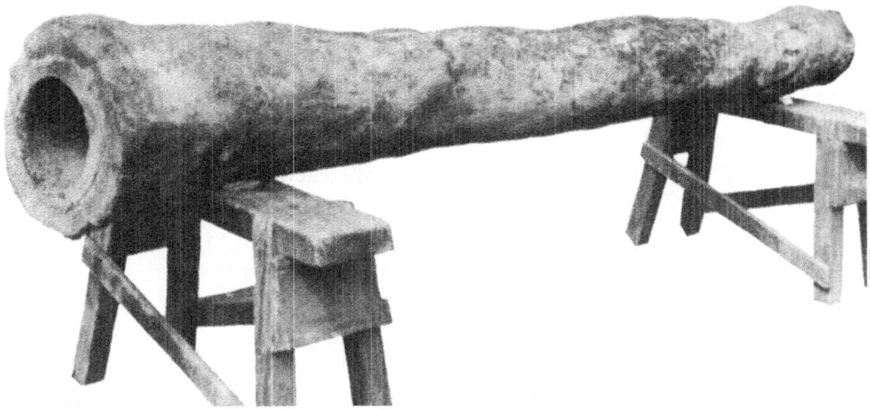

Figure 2.3 Wooden elm 'trunk' water pipe circa. 1800 (Source: Barty-King, 1992; p. 32).

Figure 2.4 Water wheel driven boring machine for wooden pipes (Source: Barty-King, 1992; p. 32)

The system which operated during this time was mainly one of storage. Most homes connected to a private company supply would have a cistern or storage tank in which they could store enough water for several days usage. This was necessitated by the practice of water companies supplying water for limited periods during the day and in most cases only on certain days of the week. Supplies would be provided to most homes having connections for typically 2 or 3 hours on about 3 days of the week.

In more prosperous homes large lead cisterns (Wylie,1959; p. 45) would be filled to supply water for other periods or, in the case of poorer families jugs, buckets or any receptacle would be filled in order to sustain a water supply during non-supply periods. The private supplies also served the very poor areas of the city. In such cases often just one tap would serve a whole street or court of ten or twelve households and storage facilities would only be possible by barrel, tub, or bucket.

WATER MAINS

The system obviously had many drawbacks, one of which was its use for firefighting. In the event of fire a runner was dispatched to the Water Company who then shut off supplies to other areas, diverting its total supply to the area of the fire. Buckets and other receptacles would then be used to put out the fire. In most cases by this time the house was probably burnt down.

The effect on the wooden water mains of an intermittent supply meant that the wooden conduits were alternately wetted and then left dry. The alternating expansion or swelling and contraction, drastically shortened the life of these wooden water mains. Even today the frequent charging and draining of water mains is regarded as bad practice.

A parliamentary commission held in 1810 received evidence from areas in the west and north-west of London that water supplies sometimes failed for two or three weeks in succession due to faulty pumps, pipes or other problems. During such times families and small communities would rely upon their neighbours for a borrowed or bartered water supply.

Iron pipes were first used for water mains around the year 1746 and the Chelsea Water Company, one of the more progressive companies in London, introduced iron water mains between 1756 and 1760. The New River Company was still employing wooden water mains at this time and continued to use wood as late as 1802 (Quennell and Quennell,1977; p. 104). It must be said that the area served by the New River Company was, however, mainly districts at a lower level and so high pressures were not experienced in the elm tree water mains under gravity flow conditions.

In 1802 the New River Company was attempting to supply new districts lying at higher levels and for this purpose employed two steam engines. To withstand the resulting pressures iron mains were used. By 1821 the New River Company is stated to have had 'some iron mains'. Iron water mains may be said to have come into general usage between 1810 and 1820 (Quennell and Quennel, 1977; p. 104). In 1817 an Act was passed requiring that after 10 years all new water mains should be of iron. However, iron mains, although an improvement, were by no means perfect. Their jointing was often very defective often resulting in enormous wastage which made intermittent supplies unavoidable where sources of supply were limited.

As more and more private water companies were formed to supply different areas of major towns, more households enjoyed the facility of a piped water supply to their homes. Initially, this was often achieved by basement storage tanks or cisterns and hand-operated pumps to supply water to the kitchen of a house or groups of houses.

FILTRATION

It became a natural progression from receiving water into the home to begin question its quality. Little attention had been devoted to the quality of the water, the provision and maintenance of a supply was until now the prime requirement.

The majority of water supplied to London over this period was from the River Thames or its tributaries. The water may have received settlement while in a storage reservoir, but there was no conscious attempt to treat it. Filtration was generally regarded as an unnecessary embellishment unless the water was exceptionally dirty. Removal of visible, fine, suspended matter by filtration was gradually introduced for river water but regarded as optional for spring water. It was only later that gravel and sand filtration were introduced as a means of removing a lot of the mud and solid matter from drinking water.

In London, in 1852, as a result of cholera epidemics, all water companies abstracting water from the River Thames were compelled to abstract above Teddington Lock to reduce sewage contamination and the filtration of all river water used for public supply was made compulsory (Quennell and Quennell, 1977: p. 37). The first successful full-scale filtration plant was designed and operated by James Simpson, engineer to the Chelsea Water Works in 1869 (Buer, 1926; pp. 229–230). It was only towards the end of the 19th century that it was discovered that harmful bacteria could be removed by the fine film of mud formed above the filtration medium.

Chapter 3

The water carriage system

WATER CARRIAGE SYSTEM

Despite the use of middens for human waste and dung piles in the streets and backyards of houses, there is much evidence that most of London during the eighteenth century was still using simple backyard privies.

THE WATER CLOSET

Inventors had been busy with plans for a water closet system as early as 1596 when Sir John Harrington, (Wylie, 1959; p. 36), who is generally credited with being the designer of the first water closet, made one for his home at Kelston, near Bath. Sir John Harrington even published a book about it in the same year and claimed that by the use of his closet 'unsavoury places may be made sweet, noisome places made wholesome, filthy places made cleanly'. The first patent application for a water closet was made in 1775 by Alexander Cumming, a watchmaker of Bond Street, London (Buer, 1926; p. 108).

Water closets were a novel feature in new houses being built at the time and initially were restricted to the houses of the very rich. Plans of large houses in 1775 showed an increase in the installation of water closets which were regarded as very desirable modern features. These devices usually discharged to a cess pool or swampy area of the garden, but sometimes flowed to a cellar or pit under the house. A common place for fitting these devices was under the stairs or in small unventilated rooms without external windows. This often caused dangerous gases, including hydrogen sulfide and methane, to accumulate as they could be drawn through the system back into the house. There were several fatalities recorded from such toxic gases.

Water closets were constantly being developed and improved by different inventors. Following the modifications and patent brought out by Alexander

Cumming, a gentleman called Joseph Bramah, who was a cabinet maker, took out a patent in 1778 for another form of valve closet. This was similar to Cumming's, the difference being in the arrangement of the valves. The pan closet dates from around 1790; being simpler to make, it was cheaper, but the container was too large to be self-cleansed by the flushing action (Stanbridge, 1976; vol. 1, p. 28).

All these types of closet were the precursors to the modern, pedestal, wash-down type of system. The general principle of all these devices was to flush away, apparently by magic, the human waste which was deposited in them. Little attention was directed as to where this waste would eventually be deposited. Whenever possible, the water closets were connected to sewerage systems if they existed.

ACCEPTANCE OF THE WATER CARRIAGE SYSTEM

Thus we have a system developing which is very similar to the system mentioned earlier which was recorded for the Royal Palace in the Minoan capital of Knossos and existed in 1600BC.

Such systems were developed in most major towns and sewage was generally conveyed to the nearest water course by sewers that were originally devised for rain-water drainage. Thus sewage was transferred from the streets to the rivers. This had a drastic effect on water supplies of the period because the rivers, which were the sources of most water supplies, were now also being increasingly polluted by the water carriage system of sewage disposal. Public attention was focused upon this problem in 1871 when the future King Edward VII caught typhoid fever while staying at a house in Yorkshire. The source of his infection was traced to bad drainage. His illness resulted in immediate efforts to improve the prevailing sanitary systems (Wylie, 1959; p. 36).

The public acceptance of the water carriage system was not immediate. There was much resistance to this method because it was initially more expensive and demanded a ready supply of water to transport the waste matter. Many people thought that the fertiliser value of sewage and excreta was being wasted by this new-fangled method of sewage disposal.

This led to a division on the subject of the general disposal of sewage. There were the champions of the conservancy system which proclaimed that night soil buckets should be collected and the sewage should be used as fertiliser on farmland, and there were those who advocated the water carriage system as being the most hygienic and cleanest way to deal with this problem. Thus, the water carriage system, as its use grew throughout society, was continually met by resistance from those who thought that conservancy was the best method of dealing with sewage.

There was a drastic increase in the population of the major cities due to migration of workers from agricultural work to the new industries. The Industrial Revolution had the effect of severely increasing the numbers of the poorer working classes. This inevitably led to water shortages and the prevalence of insanitary conditions among these groups occurred to a very much greater extent than had hitherto been experienced. Such a situation certainly helped the water carriage system to become established as the sheer ergonomics of physically removing waste matter and excreta from so many town houses became impossible. In towns, the human excreta accumulated in backyards and streets and inner courts of groups of houses and disease was rampant. Diseases were carried by flies and personal contact. Sources of drinking water, such as wells, were contaminated from cesspool overflows and sewage which percolated through the soil from privy middens.

CONTAMINATED WATER AND DISEASE

Water-borne diseases such as gastro-enteritis, dysentery, para-typhoid, and typhoid fevers were endemic in inner city areas. 'Summer diarrhoea', as it was euphemistically called, was a common occurrence and viewed with some degree of acceptance as a necessary evil. There were also more serious outbreaks of disease which claimed large numbers of lives over these periods. The main killer disease was cholera, which was first experienced in 1831 and caused over 16 000 deaths in this country. It occurred again in 1848 and the total deaths in England and Wales from this outbreak were 72 180. The last recorded outbreak of cholera was that of 1866 which claimed in excess of 14 000 lives (Stanbridge, 1976; vol. 2, pp. 42–43).

The rapid growth of manufacturing towns due to the industrial boom can be seen in the national census figures over the period. When the first census was taken in 1801 the population of the country was ten million, with three million living in towns. In 1851 the population was twenty million with ten million living in towns. This increase of seven million town dwellers over 50-year period had never occurred before in world history (Wylie,1959; p. 43).

As a result of the insatiable demand for labour by the flourishing industries, workers houses were built very quickly and insubstantially by speculative builders, landlords and owners of factories. The living conditions of the working class were almost intolerable, being densely packed into back-to-back houses, tenements or even cellar dwellings in some industrial towns.

Dr P. Boobbyer, who was the medical officer of health for Nottingham, attempted to establish a link between enteric fever and sanitation (Stanbridge, 1976; vol. 2, p. 6). By comparing places where water closets were in use with other places using pail closets and privy middens he showed that there were twice as many cases of enteric fever in the latter than in the former, and he even went further to say that there were fourteen times as many cases in houses served by privy middens. This simple comparison of disease and type of sanitation is not, however, conclusive, as there is no record of other factors such as the source of water supply for those areas under scrutiny.

Cholera outbreaks baffled the medical world at this time and some even doubted that it was contagious, although the *Lancet* recorded that it was contagious but the method of transfer was unknown (Stanbridge, 1976; vol. 2, p. 37).

In 1848-49 Dr John Snow, a Yorkshire physician, published an essay on the spread of cholera, attributing it to afflicted persons contaminating the drinking water. Although he received no official recognition in England, he was awarded a prize by the Instut de France for his essay. (Stanbridge H H, 1976, "History of Sewage Treatment in Britain" vol 2, Institute of Water Pollution Control, p. 37).

In the year 1849, Dr Budd of Bristol advanced a theory that the disease was due to living organisms from the intestinal canal of infected persons. He was also of the opinion that contaminated drinking water was subsequently the carrier of the disease. He recommended that all discharges from infected persons should be disinfected. The College of Physicians did not support his views, but despite this he reputedly controlled an outbreak of cholera at an army barracks near Bristol by this method. (Stanbridge, 1976; vol. 2, p. 37).

The confirmation that certain diseases were water-borne was proved conclusively by Snow in 1854. He established that, almost without exception, all cholera victims in an outbreak in south London had taken water from the pump at the junction of Broad Street and Cambridge Street. This pump was contaminated by the sewage from infected people (Barty-King, 1992; p. 128). These findings were supported by Farrs' statistical studies, but it was not until 1884 that Koch isolated the cholera vibrio (Barty-King, 1992; p. 128). There was thus mounting evidence of the spread of water-borne disease which changed medical opinion and gave impetus to reformers of the period.

Those against the water carriage system were quick to point out that the very introduction of this system was directly responsible for outbreaks of disease. In the densely populated urban areas disease was due in many cases to direct pollution, or eventual pollution, of water supply sources due to the water carriage system. There is evidence for both sides of this argument. Public health reformers of the time gave lurid accounts of conditions endured by the majority of the population. Dr Farr, a famous reformist renowned for his meticulous records of the time, produced a damning report upon urban living conditions in 1841. He recorded that 'the water closet throws into the sewer evacuations of the sick' and goes on to say that 'almost coincidentally with the first appearance of epidemic cholera and with the striking increase of diarrhoea in England, was the introduction into general use of the water closet system' (Stanbridge, 1976; vol. 2, p. 4). Ironically, the findings of Farr were used later by Chadwick to promote the water carriage system.

SANITARY HEALTH REFORMS

In 1838 the Poor Law Commissioners were anxious to improve the sanitary conditions in working class areas, and specifically in manufacturing towns, and asked one Edwin Chadwick to report on these matters (Stanbridge, 1976; vol. 2, p. 5). Chadwick was aided in his investigations by three doctors Dr Arnott, Dr Kay and Dr Southwood-Smith) and their reports were published as appendices to the annual reports of the Commissioners and Chadwick started preparing his classic report on the Sanitary Conditions of the Labouring Classes of Great Britain. His report was completed in 1841 and published by the Poor Law Commission in 1842 (Stanbridge, 1976; vol. 2, p. 5). The report made very grim reading. The primitive conditions he described so shocked the general public that a Royal Commission on Health of Towns was appointed as a direct result of his report to enquire further into the matter and to effect improvements wherever possible.

Edwin Chadwick was born in 1800 and was a lawyer by training, but became a practical engineer and social reformer. He devoted the greater part of his long life to combating disease and suffering and improving sanitation and living conditions for the common man. Chadwick died at the age of 94 and was knighted in recognition of the enormous benefits of his work, which even by that time had brought improvement throughout the country. He received his knighthood late in his life in 1893, a year before his death (Wylie, 1959; pp. 54–55; Anon, 1871; p. 61).

The conditions that Chadwick strove to improve were, indeed, atrocious. Chadwick's report of the sanitary conditions of the labouring population was a monumental work and drew upon evidence covering all aspects of health and social conditions, much of the evidence was gathered by Dr Southwood-Smith, Dr Arnott and Dr Kay and further information was gathered from the reports produced by Dr Farr in 1841.

PUBLIC AWARENESS OF INSANITARY CONDITIONS

Chadwick's reports resulted not only in the appointment of the Royal Commision of Health of Towns but also the passing of the Public Health Acts of 1848 and 1875 and the Sanitary Act of 1866. Chadwick acted as unofficial secretary to the Commission on Health of Towns which was appointed in 1843 and was responsible for drafting the first report published in 1844 and the second report which embodied the Commission's conclusions which was published in 1845 (Wylie, 1959; pp. 54–55; Anon, 1871; p. 61). The commission investigated the conditions prevalent in 50 towns in England, the total population of which was three million. There were also special reports on Huddersfield, Leicester, Liverpool, Nottingham, Preston and York (Stanbridge, 1976; vol. 2, p. 5).

In London a Metropolitan Sanitary Commission was appointed in 1847 to investigate conditions in the nation's capital. Their first report, published in the same year, showed that similar conditions prevailed in London as in other major cities in England. The reports described the existing deplorable conditions and clearly, the situation could not be tolerated. The public were more aware of the terrible conditions in which the majority of their countrymen were forced to live and a solution had to be found and implemented as soon as possible (Quennell and Quennell, 1977; p. 37).

INTRODUCTION OF IMPROVEMENTS

It cannot be overstated how much we owe to the efforts of Sir Edwin Chadwick. Chadwick's proposals for dealing with the problems of insanitary conditions in urban areas were simple in concept but expensive and complex to carry out. They required an integrated and rational policy by governing bodies.

Chadwick made the following recommendations (Stanbridge, 1976: p. 9):

1. The provision of an adequate and constant supply of water under pressure to every dwelling house.
2. The use of water closets in preference to privies, pail closets or earth closets.
3. The discharge of domestic waste waters direct into a sewer instead of into a cesspool.
4. The sewer to receive in addition solid refuse from the streets.
5. The sewer, instead of discharging direct into a watercourse, to convey the sewage and refuse away from the town to an agricultural area where its value as manure could be utilised.

These recommendations could be summarised as an improved water supply, adoption of the water carriage system, the provision of adequate sewers in towns and the use of sewage as manure. These proposals had far-reaching ramifications and were investigated by the Commission of Health of Towns who gave their approval and work commenced to put into practice Chadwick's proposals. The proposals were simple and very basic in their perception, but most complex to introduce on a national scale. The improvements to water supply were brought about over several years, very gradually by the introduction of various Acts of Parliament and as a result of surveys or Commissions undertaken at the time. The most pertinent of these were the Commission on Health of Towns Survey 1843–1844, the Health of Towns Association Survey 1847–48, and the Public Health act 1848.

London itself had several changes to legislation to improve the public water supply. By 1850 London water was being supplied by eight joint stock companies, operating in areas which often overlapped. There were also a number of shallow wells still in use, but these were badly polluted and there were no less than 209 public sewers discharging to the tidal portion of the River Thames.

The water companies had made vast profits from supplying water of dubious quality, often poisonous or infected, to the populace. Competition was fierce. In some cases a street would have two or more supplies and belligerent labourers employed by the water company would exert pressure upon locals to use their companies water. Plumbers made large profits by persuading customers to change from one source of supply to another (Stanbridge, 1976; vol. 2, p. 15).

In 1852, the Metropolis Water Act made the first inroads into changes to these systems. However, even by 1867 London was still without a continuous water supply and it was not until 1899 that this was corrected. Matters were made more complex by the number of different water companies involved. These water companies were eventually superseded by the Metropolis Water Act of 1902 (Stanbridge, 1976; vol. 2, p. 15), which formed the Metropolitan Water Board, a body which served the whole of London and had the responsibility of supplying water.

The gradual abandonment of the conservancy system in favour of the water carriage system of sewage disposal continued over the same period. The water carriage system was higher in initial capital outlay, but once provided, the cost of operation was much cheaper than the daily or weekly removal of night soil from the many premises within the towns. There persisted, even up to 1871, resistance to this abandonment of the immediate fertiliser value of excrement which had such great attraction at the time. The abolition of middens and substitution of water closet systems continued slowly. The general introduction of the water closet systems advanced more quickly in the south than in the north of the country; this almost certainly reflects the difference in living standards and affluence of local populations. In the northern towns it was mainly the richer people that could afford the initial cost of introducing such a system and water closets were found to be limited to the upper class districts.

SEWERS AND SEWERAGE

It was as late as 1936 under the provision of the Public Health Act, Section 47(1), that local authorities had the power to require that 'earth closets should be replaced by a water closet system if there were a sufficient water supply and sewers were available'. The general picture that emerges over this period was one of gradual replacement of the conservancy system for that of the water carriage system. The water carriage system supported by an infrastructure of sewers enabled the wastes of a community to be cheaply and easily transported. Using water as a medium and gravity as a motive force, it was possible to transport human waste and the liquid waste of a society over fairly long distances at a very low cost.

With the advent of the steam engine, steam-powered pumps enabled this process to be improved and all areas of towns could have their sewage removed. It could be transported through sewers and pumping stations to discharge to watercourses downstream of the town or to fields where its value as manure could be extracted from the pumped sewage.

The successful pumping of sewage using steam pumps was first achieved by James Vetch. Vetch had a successful career in engineering ranging from railways and mining to military engineering and eventually sewerage engineering. He designed a sewerage system for the town of Leeds and in areas where gravity could not be used to convey the sewage he pumped it through iron pipes using steam power which was very much a copy of what the water companies were doing to distribute water through water mains. The pumping of sewage by steam-powered pumps was first introduced in this country in the period of 1841–43 (Stanbridge, 1976; vol. 5, pp. 7–10).

The improvement of sewerage systems underwent similar changes over the same period as the water supply developments and as a result of Commissions and Acts of Parliament proper sewerage systems were eventually made available to most households. The legislation which enabled these changes to take place was embodied in the Commission on Health of Towns survey 1843–44, the Towns Improvement Clauses Act 1847, the Public Health Act of 1848, the Health of Towns Association Survey 1847–48, the work of the General Board of Health 1848–54, the Public Health Act's Amendment Act of 1890 and the Public Health Act 1936 (Anon, 1871; p. 54).

Where a town was served by a major watercourse or river such as London or Bristol, it was natural that all sewers gravitated to the river. This was especially the case in London where each bank of the Thames had former streams, now sewers, flowing into it which carried the sewage and dirty water from the town generally. These were most inadequate and grossly polluted the River Thames. Another unpleasant feature was due to the tidal nature of the Thames. The noxious contents of the sewer were regularly forced back up the sewer pipe during high tide periods.

This was a common feature of towns in similar situations although luckily, few had the tidal problems of the capital. In London, and later in other towns, this was overcome by considerable construction work involving the laying of interceptor sewers parallel to the river to collect all discharges flowing to the river.

The work in London was the result of a scheme prepared by J. Bazalgette, a great pioneer of such sewerage systems. His scheme involved the laying of an interceptor sewer along either bank of the River Thames to collect the waste from all the discharges to the Thames. It was a simple and innovative approach to a most difficult problem. This practice was adopted by other towns with discharges along either bank of major rivers.

In the case of London an interceptor sewer draining all London south of the Thames to a pumping station at Crossness was completed in 1862 (Figure 3.1). The sewers serving the city to the north of the Thames were collected by a similar intercepting sewer which conveyed all the sewage to storage tanks at Barking and this was completed in 1864.

By 1865 two very large storage tanks had been built at Barking for the purpose of storing sewage to allow its discharge on the outgoing tide. The sewage could now flow freely to the river, or to the storage tanks in the case of Barking and north London, and not be held up, stagnant and septic, in the local sewers during high tide periods.

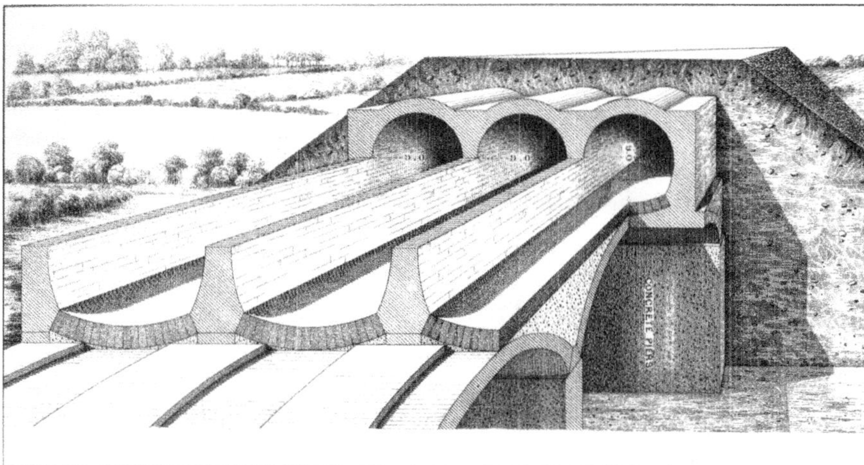

Figure 3.1 Northern outfall of Bazegette's interceptor (Source: Barty-King, 1992; p. 129)

Other towns adopted similar practices and even where there were no rivers running through the town, the installation of sewers in an urban conurbation and draining sewage to a point outside the town, became the accepted principle. The ultimate drainage was to a river, or to some area where the sewage could be used for agricultural purposes.

The result of all these changes meant that the introduction of a water carriage system made living in towns healthier and more tolerable for the working population, but the solution of the problem was not solved. By these methods the sewage effluent was transferred from the middle of the towns to the edges where vast stretches of rivers were being polluted or large areas of evil-smelling marshland were being created.

Eventually discharge to a watercourse, the sea or an estuary was considered the most apt solution, but it is only in the last 30 years that real progress has been made in dealing with pollution from these sources by developing treatment methods.

Chapter 4
Land treatment of sewage

SEWAGE UTILISATION

Chadwick himself recognised that the ultimate disposal of the collected sewage was a weakness in his overall scheme. The removal of liquid waste from the community to a suitable disposal point away from a town normally resulted in a discharge to the nearest watercourse. Gross pollution resulted from this practice and led to many injunctions being granted to prevent pollution of potential water supplies. It was recognised that sewage was a valuable source of manure which could be utilised for the growing of crops and Chadwick grasped this as a suitable method of resolving this problem.

A pioneer of 'land treatment' of sewage, as it was to be called, was James Smith. Originally Smith was the manager of a cotton mill in Scotland who also owned a small farm. He found that by using waste from the factory lavatories and also from the privies on his farm he could obtain much higher crop yields (Stanbridge, 1976; vol. 2, p. 4).

HOSE AND JET

Smith regarded the increased yields to be so successful that he set himself up as an agricultural engineer advising on irrigation of sewage on to grassland. To achieve this he adopted the idea of Vetch, using steam pumps to pump the sewage, so that it could be distributed as desired over different areas of farmland. Smith was appointed a member of the Commission on Health of Towns and contributed to their report in 1843, recommending application of sewage to land in order to profit from the growing of crops (Stanbridge, 1976; vol. 2, p. 5).

The method of distribution by hose and jet over farmland became a popular method of sewage disposal. Many entrepreneurs tried to make their fortunes by exploiting this free source of fertiliser.

The first occasion upon which land treatment as a method of sewage disposal was officially condoned or recognised, was in the Towns Improvement Clauses Act of 1847. This act stated that sewage could be disposed of into a river or into an estuary or the sea or *could be sold for agricultural purposes*, so long as no nuisance was created.

Distribution by hose and jet was undertaken in England and Scotland and large sums of money were spent on the installation of underground networks of pipes, steam pumps and jet systems to spray the sewage over the land. All this was done in the hope of making profits, but very few people succeeded in dealing with this method of treatment in an economical way.

There were many drawbacks to the system. The moving of hoses over the land damaged the crops, sewage tended to be applied to the leaves and the parts of the plant above ground rather than the root system and the volumes of sewage which had to be dealt with were very large, continuous and increased by rainfall. The volumes that the hose and jet system could cope with were limited. Another disadvantage was the untimely introduction of guano for use as a fertiliser. During the middle of the last century it was being imported into this country in vast quantities from Peru. Guano was easier to apply, and was a very rich source of nitrate. More importantly, Guano was not accompanied by the irregular and large volumes of water that were such a disadvantage with sewage (Wylie, 1959; pp. 54–55; Anon, 1871; p. 61).

SEWAGE FARMS

Despite the apparent failure of the hose and jet method, it was gradually being appreciated that land treatment was an available method of disposing of sewage from a community, especially where sea or estuarial outfalls were not available. Over a period of time from the 1870s to the start of the twentieth century, a more scientific and technical approach was given to this subject.

A committee was appointed to investigate the 'Treatment and Utilisation of Sewage' in the year 1868 and reported its findings at a meeting in Exeter in 1869. This committee was re-appointed (with a grant of £50) to report further, a final report being published in 1871. This report contains a detailed description of sewage utilisation at Breton's Farm, near Romford (Stanbridge, 1976; vol. 2, p. 9).

Experiments were carried out at Breton's Farm to determine the amount of nitrogen that could be recovered from sewage and the amounts which could be used by crops grown on sewage-irrigated land. Over a period of 3 years it was found that one-third of the nitrogen present in sewage was recoverable in the crops. At the same location it was found that phosphoric acid concentrations also increased with the application of sewage to the land. An increase in the presence of phosphoric acid by a factor of five was recorded (Stanbridge, 1976; vol. 2, p. 15).

The farm came to be used as a sewage farm in 1869 as a result of pollution of the river Rom. All of the sewage from the town of Romford was originally discharged into the river Rom downstream of the town. Being a tributary of the Thames, the Rom flowed into the Thames downstream of Dagenham.

The local sewer authority was the Romford Local Board of Health, which was constituted under the Public Health Act of 1848. As pollution increased to an intolerable extent the Court of Chancery granted an injunction restraining further discharge of sewage into the River Rom.

Initially the Board of Health attempted to dispose of the sewage by pumping it on to the lands adjacent to the outfall to the River Rom. Eventually they extended the sewage outfall to Breton's farm in the pParish of Hornchurch. The farm was purchased by the Board of Health and some 48.5 hectares was used for the 'purification and utilisation of the sewage for irrigation'.

The lower portion of the farmland was irrigated by gravitation. The rest of the farm area received sewage from two 6 kW (8 hp) steam engines which could pump the sewage to a height of about 7.5 m and discharge it into iron troughs supported on wooden trestles which conveyed the sewage to all parts of the farm (Stanbridge, 1976; vol. 5, pp. 7–10).

The population of Romford at that time was approximately 8000 of which 7000 discharged to the public sewer. The dry weather flow in 1871 was 1123 m^3 per day (Stanbridge, 1976; vol. 5, pp. 7–10). Breton's Farm is perhaps symptomatic of the state of development of sewage treatment in the country at that time, in that farmers could not be persuaded to utilise the sewage discharged from most of the major towns in the country. Because of this, health authorities had to acquire land themselves and use it to treat sewage in order to protect streams and watercourses from pollution. This involved, as in the case of Breton's Farm, the appointment of a tenant farmer and land acquisitions by the local authority. It required no ingenious engineering to distribute liquid sewage over large areas of level ground. To effectively grow crops upon the sewage irrigated land and still effect some purification of the sewage was not easy, however.

It is somewhat surprising to find the Royal Society of Arts having an involvement in sewage irrigation on farmland. This body awarded substantial prizes each year to the best managed sewage farms and this certainly helped the establishment of sewage farms in the country and led to improved management and greater crop yields but, more importantly, resulted in more adequate sewage treatment in these cases. Crude sewage could successfully be treated by these methods, and produce crops when applied at rates of between 25 and 75 m^3 per hectare per day, depending on the type of soil and sub-soil. If crop production was disregarded altogether, the rate of application could be increased to as much as 455 m^3 per hectare per day.

PROFITABILITY OF SEWAGE FARMS

The emphasis on making a profit by utilising the sewage as manure by discharging it to land was becoming increasingly difficult to substantiate. A conflict was emerging between the application of sewage to land to make a profit from growing crops (and thus using the manure content of sewage as a valuable fertiliser), and using land irrigation as a means of treating and purifying sewage to reduce river pollution.

By 1870 it was becoming more and more apparent that the profitable utilisation of sewage was impractical and wherever it was attempted, this conflict of interest between the treatment of sewage and economic crop production was most pronounced. It was becoming ever more apparent that sewage would continue to be conveyed to the outskirts of towns and pollute watercourses where it discharged untreated from outfalls, and a method had to be established to deal with this problem.

Irrespective of whether a profit could be made from growing crops, sewage could be dealt with on land without causing a public health nuisance or causing the outbreak of disease. With the increasing pollution of streams and rivers the emphasis was changing from the use of sewage as a fertiliser to the treatment of the sewage as a pollution prevention measure. Acts of Parliament had been passed to enable local authorities to purchase land, sell sewage and rent land to tenant farmers to carry out farming with sewage irrigation.

The Royal Society of Arts held a conference on Health and Sewage of Towns in 1876. The executive committee of this body reported that 'land treatment, only in certain localities where land at a suitable price could be procured, and then with favourable gradients, suitable quality of soil and sufficient area of land, could a sewage farm, if properly managed, be regarded as the best method of disposing of water carried sewage'.

They did add, however, that it was essential to bear in mind that a profit should not be looked for by the health authority establishing the sewage farm and only a moderate profit could be expected by a farmer from the utilisation of the sewage as a fertiliser.

Although large numbers of sewage farms were very uneconomical to operate on a crop sale basis and land treatment without cropping was resorted to, there were a surprising number of sewage farms which were still operating well into the twentieth century.

One of the very first sewage farms was established at Rugby and first mentioned in 1853; it was still in operation in 1909. Leicester had a sewage farm until 1965, Burton-upon-Trent until 1967, Nottingham until 1956 and Breton's Farm at Romford was still operating as a sewage farm until 1969 (Anon, 1871; p. 54).

LAND AREA REQUIREMENTS

The great attraction of applying sewage, or settled sewage, to beds of different media for treatment, lay in the ability to apply greater volumes to smaller land areas.

In 1864 there were seven towns in England practising sewage farming and this figure gradually increased until in 1876 approximately 60 towns were using sewage farming or land treatment of some type. It is suspected that the number of towns undertaking land treatment which were not necessarily associated with successful farming enterprises was greater than this. The use of land treatment and sewage farming to produce better sewage effluents that could be more safely discharged to inland watercourses was adopted with great commitment by the Royal Commissions on Sewage Disposal. This acceptance was reflected by the Local Government Boards which were responsible for government loans to local authorities who wished to spend money on sewage treatment.

A pattern had been established by the Local Government Boards such that when a town was about to embark on a sewage treatment project, the board would insist on certain amounts of land being purchased relative to the population whose sewage was to be treated. The relative populations to one hectare of land were well established. Where broad irrigation was practised, one hectare of land was to be provided for every 740 people. Where intermittent downward filtration was to be the method of application of sewage to land, the sewage from 2470 people could be treated on one hectare of land. For cases where chemical treatment (see later in this chapter) was to be followed by broad irrigation, the same figure was used. Where chemical treatment was followed by intermittent downward filtration, only one hectare of land was required for every 4940 people served. This tends to illustrate the Local Government Board's confidence in chemical treatment followed by intermittent downward filtration. These were the two best methods of sewage treatment available at that time.

This insistence on large tracts of land being made available for sewage treatment was often an impediment to pollution prevention, since in many cases there were long delays while the local council purchased sufficient land. This was particularly so in the northern industrial areas and manufacturing districts around Manchester, Leeds and Sheffield and in Yorkshire and Lancashire generally. The Local Government Board insisting on fixed levels of land area per population, had to stipulate that very large land areas would be required for these northern towns as they were forever increasing in size and population. There was insufficient land available for sewage farming or land disposal in these areas and its cost would have been prohibitive.

This central government attitude persisted, even after the advent of better sewage treatment processes, which did not require such large areas of land.

As a result of a survey made in 1870 by the Committee on the Treatment and Utilisation of Sewage, it was found that of 189 towns and districts having systems of sewerage, 143 discharged their sewage without any treatment whatsoever and in 17 cases the sewage was simply filtered before discharge. In seven instances it was chemically treated and in 17 cases, irrigation was used. At five locations this system of disposal included more than one of the above methods.

INCREASED FLOWS

Land treatment of sewage was made more difficult due to high volumes and continuous flows. In almost all cases the volume of sewage discharged was much greater than the volume of water supplied to the community. In many cases this difference in volume was attributable to combined drainage systems and the fact that rainfall swelled the sewage flows. Information produced by the Report of the Committee on the Treatment and Utilisation of Sewage 1869 to 1870 also drew attention to the fact that infiltration water and ground-water must be augmenting sewage flows from communities which had sewers.

Many examples of this were apparent in their report of 1870. Leamington Spa, for example, was reported to have a population of 20 000, a water supply of 1800 m^3 per day and a sewage discharge (which incidentally was discharged untreated direct to the river Leam) of 4550 m^3 per day. Another example cited was Torquay, with a similar population to that of Leamington Spa (20 000), a water supply of 136 L per head per day which gave a predicted sewage flow of 2770 m^3 per day. The actual sewage discharge from Torquay, however, amounted to 7270 m^3 per day. The point made by the report of 1870 was that in many cases these larger flows made pumping of the sewage too expensive for the local population and thus discharge direct to a watercourse or to the sea was the only economic solution. To apply the sewage to land in so many cases meant pumping all or part of the sewage flow.

CHEMICAL TREATMENT

Efforts were made to reduce the land area which was necessary for land treatment of sewage by pre-treatment of the sewage before its application to land. The most common form of pre-treatment was settlement in tanks of varying sizes proportional to the sewage flow. For quite short retention times only coarse settlement resulted. The overall effect, however, was to reduce the load exerted by the settled sewage and this did enable smaller land areas to be used for conventional land treatment or sewage farming. The settled matter was usually collected and distributed on to farmland as a fertiliser.

There were many experiments with chemical treatments, the most common being the use of lime as a precipitant aid at this first stage of sewage treatment. On page 139 of the report of 1872 on the Treatment and Utilisation of Sewage, there is a most interesting description of an experiment carried out by a General Scott. He attempted to treat the sewage of Ealing by lime precipitation and consequently discharged a much clarified, inoffensive effluent.

The lime and some ground clay were added to the foul sewer upstream of a sedimentation tank to allow sufficient mixing. The dosage rate was '10 cwt of lime and 5 cwt of clay to every 400 000 gallons of sewage'. When sufficient sludge had accumulated in the settlement tank, it was drawn off, placed in a kiln and burnt by intense heat and then ground into cement. The committee inspecting this process reported upon it with some enthusiasm as 'a convenient way of disposing of sewage sludge and producing a useful by-product, namely cement'. The costs, however, were too expensive and the sludge was far too wet even after passage through a hydraulic press to reduce the water content enough for efficient drying in a kiln. Due to technical problems, the fuel costs of heating the kiln and the chemical costs, this was not a profitable process.

The addition of lime and settlement of solid matter in crude sewage was made popular by entrepreneurs who intended to make their fortunes from selling the precipitated organic matter as fertiliser.

In Coventry, in 1874, as a result of an injunction issued by the Court of Chancery, the Corporation had to make some effort to treat the sewage from the town, which was discharged untreated into the River Sherbourne (Royal Commission on Metropolitan Sewage Discharge, 1884; p. 250). Coventry's first sewage treatment works was constructed at Whitley and consisted of several large tanks and a lime-dosing process. This was a private enterprise attempting to make a profit from selling the accumulated sludge. In 1876 the Corporation took over the treatment plant since it was proving to be a commercial failure (Tilley, 1970; p. 4).

Where the profit motive could be disregarded, settlement and even chemical treatment with lime was effective to reduce the pollution load for the methods of land treatment employed at the time. In many cases settlement and lime precipitation were adopted as an alternative to land treatment where the purchase of land would have been very expensive or where efforts were being made to reduce the pollutant load on the receiving rivers.

Birmingham had a lime precipitation process and a press house to deal with the sludge, which was formed into cakes of approximately 45% moisture. Difficulties were experienced in disposing of the pressed-cake since local farmers were reluctant to take it if it were too wet or contained too much lime, and they only needed it at certain times of the year and production was, of course, continuous (Royal Commission on Metropolitan Sewage Discharge, 1884; p. 17).

There were also plans to add lime to London sewage at the Barking Creek and Crossness outfalls. Several engineers competed for this proposed scheme and many calculations were made regarding its possible effects. In 1882 the daily flow to Barking was 305 ML and that to Crossness was 286 ML, a total of 591 ML per day (130 million gallons per day) of sewage from London. An estimate was made that this would produce approximately 124 tonnes of dry solids per day at Barking and 105.5 tonnes at Crossness. The total solids with 90% moisture came to 1209 tonnes per day, if the sludge were pressed and a 45% moisture cake resulted, this volume would be reduced to 418 tonnes of press-cake per day. (Royal Commission on Metropolitan Sewage Discharge, 1884, p. 18).

The plan proposed by Sir J. W. Bazelgette was that sludge would then be disposed of by barges which would convey this waste material out to sea. His plans had a landing stage for barges included as part of the overall scheme (Stanbridge, 1976; vol. 6, p. 26).

As water supplies were improved and urban populations increased, the water usage, and therefore volumes of sewage, became greater. These sewage flows were also much more dilute. This had the effect of making the cost of chemical precipitation in most cases prohibitive. Thus, chemical treatment of crude sewage tended to decline, but the primary settlement stage remained as a most important method of reducing organic load by a simple physical process of natural settlement.

Chapter 5

Contact beds and biological filters

FILTRATION EXPERIMENTS

There were two recognised methods of applying liquid sewage to farmland and distributing it over the surface to percolate through the soil. These were broad irrigation, also known as surface irrigation, and intermittent downward filtration. The latter was pioneered by Sir Edward Frankland.

With broad irrigation sewage was directed continuously over the top layer of the soil and this was the only treatment process. As it was necessary to interrupt the surface flow for the purposes of planting and harvesting crops, an area of land could only be used intermittently on this basis.

Sir Edward Frankland realised that by furnishing under drains, located at some depth below the ground, the land could be used as a physical filter, especially if aeration of the soil could be achieved by adequate under drainage and the sewage could be applied intermittently. This process was therefore named intermittent, downward filtration (Stanbridge, 1976; vol. 5, p. 30).

Frankland's experiments were the first to investigate land treatment from a scientific aspect. He conducted a series of laboratory experiments in 1870 for the Second Royal Commission on River Pollution. Frankland did porosity experiments with a variety of different soil types using a technique whereby a layer of gravel was placed in a glass cylinder, followed by a layer of the soil to be tested, extending to a depth of 1.5 m, and this was topped with a small layer of fine sand to prevent solids choking the voids in the soil to be tested. To ensure that air could be introduced into the soil and thus prevent blockage and ponding, a glass tube was passed down the centre of the cylinder to within 700 mm of the base to allow free access of air (Stanbridge, 1976; vol. 5, p. 29).

Frankland carried out experiments on five different types of soil which he obtained from different localities and applied to each filter a dose of equal quantities of London crude sewage, twice a day, for about four months. He found that the effluents draining from the tubes were clear, almost colourless and odourless and contained considerable quantities of nitrate. Frankland concluded that porosity and fineness of division of the soil particles were important factors in application of sewage to land and were more important than the chemical composition of the soil.

Frankland's work was further extended by J. Bailey-Denton who was a land drainage engineer. He appreciated the main disadvantage of Frankland's procedure was that of cropping and harvesting the treated land. Bailey-Denton had observed that where land was under drained, the water flowing from these under drains was invariably clear, that water or sewage flowed rapidly through the soil to the under drains under most conditions and also that plant growth flourished where under drainage was carried out.

Bailey-Denton adopted the procedure of forming ridges and furrows in the land to be treated by downward filtration. In this way crops could be freely grown and harvested on the top area of the ridges, while at the same time, sewage could flow along the furrows. He conducted experiments proving that sewage from a population of 400 people could be treated on one hectare of land if aeration by under drainage to a depth of 1.8 m was carried out, and the sewage was applied in precise quantities intermittently.

Downward filtration was used at Merthyr Tydfil in 1871. As a result of an injunction granted against Merthyr Tydfil in 1868 for pollution of the River Taff, it was necessary to quickly provide some means of sewage treatment. Bailey-Denton applied his methods of downward filtration to treat the sewage from Merthyr Tydfil and from January 1871 the sewage from a population of 20 000 was treated on 8.1 hectares of land. By 1872, a dosage rate of 1240 people per hectare was regarded as the most practicable and this method of sewage treatment was employed for some time at this location (Stanbridge, 1976; vol. 5, p. 33).

Bailey-Denton continued successfully to promote this method of sewage treatment at several other locations. In most cases he used intermittent downward filtration in combination with broad irrigation and in some cases, a small area planted with osiers for treating storm sewage. He scaled up his experiments and applied them to four plots of equal area dosed in rotation. Each plot was dosed with sewage for 6 hours a day and from this he concluded that 2 hectares of land should be sufficient to treat the sewage from a town of 10 000 inhabitants. This method of provision of underground drainage some 1 m below the surface in conjunction with a porous soil condition enabled more sewage to be treated per unit area of land (Stanbridge, 1976; vol. 6, p. 37).

The work of Frankland and Bailey-Denton was directed towards the physical filtration and porosity of the soil to be used for land treatment of sewage. Although Frankland had observed a conversion of ammonia to nitrate resulting from passage of sewage through the soil, his activities were directed mainly towards the maximum application of sewage to land area. This filtering process strongly influenced early attempts at treating sewage artificially.

Edward Frankland continued to conduct experiments with downward filtration of sewage through different, porous, soils and gravels. One of his experimental filters using coarse porous gravel was found to be most successful, producing a well nitrified effluent from a filter which exhibited no clogging or ponding after a 4-month period of operation. Following this successful operation of an oxidising filtration device, it was thought that a filter bed could be prepared which would be more porous than natural soil and sub-soil, and would be capable of oxidation throughout its whole depth. Over this period between 1880 and 1890, experiments with filtration were carried out at several locations around the country (Lechevalier and Solotorovsky, 1974; pp. 267–271).

MICROBIOLOGY OF SOILS

The role of micro-organisms in this process was not investigated by these pioneers. The work of Frankland and Bailey-Denton was directed towards the physical filtration and porosity of the soil to be used for land treatment of sewage. Although Frankland had observed a conversion of ammonia to nitrate resulting from the passage of sewage through the soil, his activities were directed towards the maximum application of sewage to a land area. The effect of microbial action was not considered.

Until the invention of the microscope there was very little knowledge of anything which could not be observed by the naked eye. The invention of the microscope is attributed to Anthony van Leeuwenhoek, a Dutch spectacles maker, who lived most of his life in Delft, Holland. He died at the age of 91 in 1723, by which time only very few people were even aware of the microscopic world. From this time onwards, however, such knowledge was increasing (Lechevalier and Solotorovsky, 1974; p. 268).

The eighteenth-century microscopes were primitive instruments and at best, gave magnifications of 200. As a result of the work of Pasteur and Koch, there was much interest in microscopy and microbiology towards the end of the Seventeenth century. With regard to soil microbiology it had long been known that nitrates tended to accumulate when organic matter was decomposed. Davy, in 1814, expressed an opinion that nitrates were formed from atmospheric oxygen in combination with ammonia in the soil.

In 1877, Theophile Schloesing, an agricultural chemistry professor at the French National Institute of Agronomy, in conjunction with Achille Müntz, published a paper which dealt in some depth with the oxidation of ammonia by soil bacteria. Their method of investigating this was similar to that of Frankland in that they passed sewage effluent through columns of soil in glass tubes, but they measured the ammonia present in the influent and the ammonia and nitrate present in the effluent.

Upon commencing the experiment there was no change in the ammonia as the sewage passed through the columns of soil. Some days later, however, the ammonia was noticed to be diminishing while the nitrate increased and eventually they established that all the ammonia was being converted to nitrate as the liquid passed through their columns. The experiments commenced with sterilised soil and half way through the experiment, when they were producing nitrate, they sterilised the soil again by the use of chloroform. Immediately after the application of chloroform the nitrification ceased and ammonia was again discharged from the base of the columns. They concluded that some living organism, killed by the chloroform, was responsible for the oxidation of ammonia (Stanbridge, 1976; vol. 5, p. 15).

An English soil chemist, called Warrington, who between 1877 and 1891 worked on nitrification within soil, later demonstrated that the conversion of ammonia to nitrate by soil bacteria was done in two stages. Warrington discovered that there was conversion of ammonia, initially to nitrite, which in turn was transformed into nitrate (Stanbridge, 1976; vol. 5, p. 15). Warrington, however, was not able to isolate the responsible micro-organisms. This was achieved by Winogradsky, a Russian, while working in Zurich in 1890. In fact Winogradsky had isolated only the organism responsible for oxidation to nitrite and this is explained in two papers he produced at the time.

In 1891, Warrington described in the *Journal of the Chemical Society* the results of his experiments between 1880 and 1881, which had revealed the existence of an organism that converted nitrites to nitrates, but could not apparently oxidise ammonia (Bulloch, 1936; pp. 24–26).

In a paper presented to the French Academy of Sciences in1891, Winogradsky explained that his previously isolated organism which oxidised ammonia to nitrite was now accompanied by his isolation of the second organism which converted nitrite to fully oxidised nitrate. These were respectively known as nitrosomonas and nitrobacter. It was not until several years later, however, that the presence of these organisms became regarded to be essential for the efficient treatment of sewage, nor were they even considered as useful indicators that treatment was effective (Bulloch, 1936; pp. 24–26).

Whereas contemporary sewage treatment is recognised as a predominantly biological process, little was realised, or advantage taken, of biological use of micro-organisms in these early days of sewage treatment. Attention was directed mainly towards physical filtration, whether through soil or some artificial medium.

Efficient sewage treatment at this time was regarded as that which did not cause gross pollution of the receiving watercourse and was judged mainly by visual examination of the effluent waters.

LAWRENCE EXPERIMENT STATION

It was in America, at the Lawrence Experiment Station in Massachusetts, where a real breakthrough occurred in biological filtration. In 1887 the Massachusetts State Board of Health investigated methods of treating sewage by passage through soils to oxidise the organic matter. They constructed several beds each of 1.52 m depth and a gravel filter, which was 1.12 m deep. It was found that much greater quantities of sewage could be treated on the gravel filter with the coarser media than those with finer media and, moreover, they attributed the changes that took place to be essentially chemical changes rather than those due to straight filtration.

Following the success of the gravel filter in July 1891, a larger gravel filter was constructed and sewage was applied at frequencies of approximately 20 minutes. This was the first 'trickling filter' or 'bacteria bed' and the most important fact that emerged from their experiments was the conclusion, which still holds today, that the slow movement of the liquid in thin films over the surface of the stones in contact with air was the most successful method (Stanbridge, 1976; vol. 5, p. 17).

A report of this research was prepared by Hiram F Mills and published in 1890. The results of the Lawrence experiments especially the effect of the use of gravel, which was almost accidental as a choice of media, had a significant impact upon workers in Britain.

THE CONTACT BED

W. Santo Crimp and W. J. Dibdin, both employees of London County Council and also J. Corbett, Borough Engineer of Salford, were all influenced by Mills' report. Crimp and Dibdin carried out experiments completely independently from Corbett and their work led to the introduction of the contact bed. The experiments carried out by Corbett resulted in the biological filter. Thus, the work at the Lawrence Experiment Station in America had a direct bearing on sewage treatment developments in the United Kingdom.

The principle of using beds of inert material with excellent drainage properties was considered to be most advantageous. Such filters would require a much smaller land area for treatment and would be of particular value where the type of soil local to the sewage treatment farm was unsuitable for conventional land treatment due to poor porosity.

The first contact beds were rectangular tanks filled with inert material such as burnt clay; these were then operated on a fill and draw basis. By means of a valve at the base of such a contact bed, sewage could be held in the bed for differing periods of time to establish an optimum contact with the media.

After being held in a contact bed for a period of time, the outlet valve was opened and the sewage allowed to flow away for land treatment or discharge direct to the watercourse. The empty contact bed was allowed to stand for several hours during which time the deposits on the media in the bed became atmospherically oxidised (Stanbridge, 1976; vol. 5, p. 2).

Initially it was the practice to treat crude sewage by this method, but problems were experienced with high solids levels which tended to blind and clog the contact bed. It was eventually established that treatment of sewage after settlement in conventional settlement tanks, septic tanks, or following chemical treatment, was more effective than treating crude sewage.

In 1893 Santo Crimp was working on contact beds at London's Beckton sewage works, investigating the treatment of settled sewage. Santo Crimp had been appointed by the London County Council in 1890 to manage sewerage and sewage treatment works in the north London area. Crimp experimented with various types of media within his contact beds. Many different types of ballast were used ranging from coke and burnt clay to polarite (a type of Dolomite limestone). Crimp found the best media to be coke breeze. This was probably due to the large surface area afforded by this material (Bolton and Klein, 1971; pp. 81–82).

Crimp explained his reasons for operating a fill and draw system. With the outlet valve at the base of the tank open, any sewage introduced would pass too rapidly through the contact bed. He was also experiencing much difficulty distributing sewage over the surface of the bed, thus by closing the base valve and flooding the bed completely, contact of the sewage with the entire volume of media was assured. By charging the contact bed fully and leaving it standing for one hour, the bed was then emptied and permitted to stand empty for 4 hours during which time oxidisation of organic matter within the bed took place. This method was less expensive than constructing a distribution channel system on the bed's surface. After the 4-hour rest period, the process was repeated (Stanbridge, 1976; vol. 6, p. 26).

Purification was very similar to that occurring during land treatment, but advantage could be taken of the smaller land area required to treat similar volumes of domestic sewage. The process was really a direct development from intermittent downward filtration discussed earlier.

Two stage contact beds were found to be more effective than single beds, with the contents of one contact bed flowing onto a second and receiving treatment in both. In fact, this method was regarded as a means of achieving a high degree of purification and was adopted at Manchester and Sheffield as well as Blackburn, Swindon and Aldershot.

J. W. Dibdin investigated the performance of the contact beds at Beckton while he was a chemist to the London County Council. He found that by filling the bed for one hour and having a 4-hour dwell time when empty, that a 75% reduction in oxidisable organic matter could be attained. Following his experiments, Dibdin developed and publicised the process describing it basically as a concentrated and accelerated type of land treatment which was most effective and economical as large volumes of sewage could be treated on relatively small land areas.

In an effort to produce a self-cleansing contact bed he introduced a slate bed, the media within this bed being horizontally stacked slates with spacers. This enabled easier flushing of deposited and hopefully digested solids from the contact bed, after leaving the bed empty for several hours. This practice would appear to be in direct conflict to the later findings of McGowan in Manchester. Slate contact beds were installed at several sewage treatment works, some were in operation at Faversham until 1973. (Stanbridge, 1976; vol. 6, p. 30).

Throughout the 1890s this method of sewage treatment became more popular and was employed at Leicester, Sheffield, Hampton, Manchester and Leeds. In many cases secondary or sometimes three-stage contact bed systems were operated. Each contact bed application produced progressively better oxidation of organic matter (Stanbridge, 1976; vol. 6, p. 29).

Over a period of five years from 1895 to 1899, G. McGowan treated Manchester sewage on experimental contact beds. Eventually this resulted in a full-scale treatment of the sewage by this method. In 1895 two beds were constructed, filled with differing graded media. In 1899 further contact beds were constructed to treat screened or settled sewage, or septic tank effluent, and the experiments were arranged such that this influent could be treated on single-stage, two-stage or three-stage contact beds.

The results of these experiments were submitted by McGowan to the Royal Commission on sewage disposal. It is interesting to note that McGowan concluded that as much suspended matter as possible should be removed from the sewage prior to application on contact beds, but more importantly, that the remaining solids within a contact bed should be retained. This is of great importance since it indicates recognition of the fact that the solids produced within a contact bed were biomass that was essential to the bed's operation. The earlier practice of washing such matter from the bed only tended to reduce the efficiency of the treatment. Although he did not perhaps fully appreciate this at the time, the retention of active biomass drastically improves treatment. This is now a basic principle and the importance of such retention was not fully realised until the work of Ardern and Lockett and the development of the activated sludge process which is discussed later.

Three-stage Ccontact beds were installed at Leicester sewage works over a period of 1901 to 1909. During this period Leicester had a total contact bed area of 4.86 ha. They were most sophisticated units and worthy of mention. Different graded media was used throughout the bed.

The problem of even distribution over and through the bed, was overcome at Leicester by distributing the incoming sewage via a wooden trough placed on the bed surface which had perforated cast iron pipes at right angles to its length. The wooden carriers complete with the cast iron perforated arms were moved to a different position on the bed's surface at least once a week. The beds employed automatic airlock inlets and timed discharge siphons which operated on an 8-hour cycle (Stanbridge, 1976; vol. 6, p. 33). This meant that the contact bed would discharge intermittently when the siphon was activated at intervals of 8 hours. Thus, operation of these contact beds was almost completely automatic.

BACTERIA BED

It is not possible to say which came first, the contact bed or the bacteria bed, since developments were continuing at the same time for both of these devices.

The early work of Frankland and the publication of the experimental work undertaken at the Lawrence Experiment Station in the USA led to a vigorous interest in sewage treatment by filtration through various media as an alternative to conventional land treatment. Experiments were carried out under the auspices of various local authorities burdened with the problem of treating ever increasing volumes of sewage. Development was due predominantly to the interest and enthusiasm of those involved.

The development of the bacteria bed as we know it today is usually attributed to Joseph Corbett. If he was not the first, he was certainly one of the first, to establish and promote this treatment technique. Corbett was quick to appreciate the need for efficient distribution of sewage over the surface of a biological filter bed and the vital importance of aeration. In the early days of development there was little to choose from the point of view of efficiency between the contact bed and the biological filter and the early adoption of contact beds at sewage works, such as Manchester, led to some neglect of the bacteria bed as an alternative means of treatment.

In 1893, when Corbett was Borough Engineer of Salford, he began experiments on small-scale biological filters. These filters, six in number, were constructed to treat sewage which had received chemical treatment. Each filter had three horizontal layers with a ventilated floor between each. Distribution was by means of fixed wooden troughs set in the top sand layer of each bed. Three different kinds of media were used; two beds were filled with gravel and sand, two with coke and two with cinders. All of the beds had a top layer of 800 mm of sand. Each filter was almost 12 m^2 in surface area and more than 1.5 m deep.

Corbett carried out further investigations into the importance of aeration within a bed by eliminating several of his ventilating floors. Thus, one bed would be totally full of graded media, another had a separate stage half way down, another with two layers, another with three and one with four layers, each having a ventilated floor between each layer.

Upon finding there was little difference in the efficiency of these filters, Corbett thought it was unnecessary to have the separate layers with a separating air space in the depth of the filter and a bed could be filled entirely with graded media but be well ventilated from the base of the bed (Alan, 1929; p. 14).

Over a period of 5 years, 1893 to 1898, Corbett established methods of ventilating bacteria beds, experimented with different types of distribution, bed construction, including raised tiled floors, different media, application rates and many other aspects. The beds were finally dismantled in 1898. He even produced the first fixed-jet distributor (circa. 1896) during his period of experimentation at Salford.

At the turn of the century it was gradually being appreciated that distribution on the surface of the bed in an even manner, and good ventilation and drainage by the grading of the medium, was advantageous. Experiments were conducted at many places to investigate and improve these aspects.

Providing a regular, even flow of sewage over the entire surface of a biological filter was much improved by the adoption of a self-propelled distributor. The first of its type was attributed to F. P. Candy who patented such a device in 1898. This relied on a reaction jet principle, the same principle that is applied to lawn sprinklers today. The sewage to be treated was fed from above and flowed by gravity along radial distributor pipes which rotated from the centre of a circular bacteria bed.

This patent by Candy followed work done initially by T. Caink who patented a similar device in 1897. The Caink device consisted of a central, vertical feed pipe, which discharged into a bell mouth on a rotating annular chamber to which a single arm was connected. This arm was supported at the extent of its radius by runners travelling around a circular track laid on the filter surface.

A 6.4-m diameter rotating distributor was installed at Reigate in 1899. Other similar types of distributor were installed later at Southampton and other locations where they became known as Candy–Caink distributors.

Whittaker and Bryant were the first to introduce a central column distributor in 1898 and from this time many different methods of distribution were evolved. Central column distribution was more attractive since the sewage flow was delivered from below through a central column to the rotating arms and replaced the 'feeding from above' principle (Figure 5.1). There were problems, however, associated with the need for reliable water seals in such a rotating system.

Figure 5.1 Rotating arms distribute settled sewage evenly onto the surface of a bacteria bed (Source: Barty-King, 1992; p. 13)

In 1905, McGowan and Frye compared the effectiveness of contact beds and biological filters at Accrington for the Royal Commission on Sewage Disposal from 1905 to 1907. Their findings were essentially that biological filters were superior in that they could treat greater volumes of sewage to a higher degree of purification for longer periods of time with less land areas than that required for contact beds. It was found that the filters could treat nearly twice the amount of sewage as contact beds of the same volume and produce a better quality effluent.

Contact beds came to be regarded as inferior to biological filters which generally replaced them in usage throughout the country. Biological filtration became the predominant method of sewage treatment rivalled only by the activated sludge process.

THE BIOLOGIC WHEEL

A process which could be regarded as a complete reversal of the 'Haworth Paddle System' was reported in an experimental stage in America in 1929. In Hammond, Indiana, under a patent held by A. T. Maltby, a company called Page & Jones Chemical Co. was experimenting with a twelve-bladed paddle wheel to achieve sewage treatment. A flow of sewage was employed for rotating the paddle wheel, alternately submerging and exposing the blades. On the surface of the blades, biological growth accumulated in which aerobic organisms were prolific. The process involved recirculation and by this method retention times from 4 to 6 hours were made possible.

With a contact period of 6 hours and the number of blades increased to 36 per paddle wheel, there was a reported 5-day biochemical oxygen demand (BOD^5) reduction of between 60 and 80%. The company claimed the land area required for this process would be approximately one-tenth of that required for trickling filters. The power costs were claimed to be low compared with activated sludge processes and there were also claims that nitrification could be accomplished by this method.

This process was novel since biological growths normally occurring on the surface of bacteria bed media were being developed on the surface of the paddle wheels. Due to the alternating exposure to air and immersion in nutrient the treatment was analogous to sewage trickling over bacteria bed media. (Doman, 1929).

CONTACT FILTER WITH PARTIALLY SUBMERGED ROTATING PLATES

This process is very similar to the biologic wheel and was a precursor to the rotating biological contactor (RBC) process widely used today. Doman constructed a prototype rotating disc sewage treatment unit in Connecticut, America, in March 1925. This plant was constructed and operated for a period of some 20 weeks.

A sewage flow of 12.7 L per hour was treated; it had a contact tank retention period of 1.7 hours and the plates were rotated at one revolution per minute. The contact tank was 21.36 L in capacity and was served by 14 discs, each 35 cm in diameter, spaced at 2.5 cm centres. A 373 W motor was provided for continuously revolving the plates and this motor also operated the feed system to the unit, a bucket elevator.

This experimental unit was treating septic tank effluent in what was essentially an aerobic system. Initial results were inconclusive. Increasing the retention time to 6 hours and decreasing the revolutionary speed to half a revolution per minute gave more promising results (Kershaw, 1977). This is the first known record of a rotating biological disc unit.

Mention should be made of the Association of Managers of Sewage Disposal Works, which was founded in November 1901 as the result of a letter sent by Daniel Ledson to a publication called the *Sanitary Record*. This association was formed to promote the advancement of scientific and practical sewage treatment. By holding meetings, presenting and publishing papers, a vast amount of knowledge was made available. Such interchange of ideas and information was so essential for the rapid advancements that were made at the turn of the century in this field. This association later became the Institute of Sewage Purification, then the Institute of Water Pollution Control and later the Chartered Institute of Water and Environmental Management we recognise today.

Chapter 6

The activated sludge process

AERATION EXPERIMENTS

With the realisation that oxidation was essential for the purification of sewage many attempts were made to hasten this process by direct air injection, or the forcing of air into sewage. In 1882 Dr Angus Smith reported that aerated sewage, in all cases, caused putrefaction to be delayed and nitrates to be formed more readily in the aerated rather than in the non-aerated samples.

The effectiveness of direct aeration of sewage was investigated at the Lawrence Experiment Station from 1891 to 1898. Similar experiments were undertaken by Locock in 1892 in England, but these experiments were upon the aeration of biological filters and not direct aeration of sewage held in tanks.

In 1897 Dr Fowler carried out experiments on the aeration of sewage effluents at the Davyhulme sewage works of Manchester Corporation. Fowler eventually obtained clear liquid and a deposit which settled rapidly. The solids he produced were to become known as 'active sludge'. The significance of the suspended matter produced following aeration was not appreciated at that time. Fowler was attempting to liquefy or gasify the total insoluble matter present in the sewage (Bolton and Klein, 1971; pp. 181–184).

In 1912 Dr Fowler was still based at Manchester's Davyhulme sewage works. In the same year E. M. Mumford was working at Manchester University when he discovered an aerobic organism which caused iron to 'floc out' from colliery water. Mumford designated this organism 'M7'. E. M. Mumford with the help of E. Ardern then tried to simulate the coagulation which occurs within a bacteria bed causing biomass to be deposited on the surface of the media. They inoculated a culture of M7 in the presence of a small amount of ferric salt into settled sewage. Since the organism (M7) was known to be aerobic, air was blown through the mixture for some 6 hours at a sufficient rate to maintain adequate circulation within the vessel and after this time a clarified effluent was produced.

AMERICAN EXPERIMENTS

Dr Fowler, while visiting the Lawrence Experiment Station in the United States in the same year, observed similar experiments being carried out by Clark and Gage on the aeration of sewage, but with the addition of green algae. The object of these experiments, which were commenced in April 1912, was to examine the effectiveness of aeration in combination with green algae as a preliminary treatment of sewage prior to application on bacteria beds.

The biological purification of sewage carried out in the presence of sufficient air, bacteria and other micro-organisms results in the coagulation and flocculation of colloids and pseudo-colloids as well as oxidation of carbonaceous matter and oxidation of ammonia (Bolton and Klein, 1971; p. 37).

The motivation for the Clark and Gage experiments was that if coagulation and carbonaceous oxidation could be conducted before application of the sewage to the bed, there would be less deposit within the bed of coagulated material. There would thus be less chance of blinding biological filters and therefore a higher rate of application would be possible.

WORK AT MANCHESTER

Upon his return from America, Dr Fowler instigated research into the direct aeration of sewage in order to effect complete nitrification. The presence in Manchester of E. M. Mumford, E. Ardern, Dr Fowler and later W. T. Lockett was most fortunate and led to the discovery of the activated sludge process (Institute of Sewage Purification, 1954a; pp. 174–186).

Edward Ardern was already resident chemist at the Davyhulme Sewage Treatment works, Manchester, and had been responsible with Dr Fowler for the installation and operation of the extensive contact bed construction at this works. William T. Lockett joined the staff in the laboratory at Davyhulme, initially as an assistant chemist, later to become a senior assistant and research chemist, a most valuable asset to the developments that took place at Davyhulme over the following years.

THE DISCOVERY OF ACTIVATED SLUDGE

The discovery of the activated sludge process is attributed to Ardern and Lockett in 1913. Their findings were first published in a paper entitled 'Experiments on the oxidation of sewage without the aid of filters', which was presented to the Society of the Chemical Industry on 3 April 1914 (Bolton and Klein, 1971; pp. 181–184).

Essentially, the findings of Ardern and Lockett were that if air were passed through settled sewage for a period of time, eventually solid matter would develop within the liquor. If this solid matter was retained and mixed with more settled sewage and aerated in the same way, the periods of aeration needed to achieve oxidation and nitrification of the liquor became shorter and shorter. The solids separated from treated sewage, when retained for re-use with a subsequent volume of untreated sewage, enabled organic matter to be oxidised in a period of 24 hours or less instead of several weeks.

They quite correctly attributed the increased treatment to the presence of the solid matter or sludge which they were retaining and re-using. As they say in their paper 'for reference purposes and failing a better term, the deposited solids resulting from the complete oxidation of sewage had been designated "activated sludge"'.

As a result of preliminary experiments Ardern and Lockett observed the following conditions.

(a) The alkalinity of the influent sewage should be equal to, or in excess of, the acidity due to nitric acid formation from nitrification of ammonium salts. In some cases addition of alkali may be necessary prior to complete nitrification.
(b) It was essential that activated sludge should be kept in intimate contact with the sewage during aeration.
(c) The relationship between the volume of activated sludge and the volume of sewage to be treated was of great significance, especially with regard to the rate of nitrification (Bolton and Klein, 1971; pp. 181–184).

The laboratory experiments carried out by Ardern and Lockett were repeated on a pilot scale, then a larger scale plant at Davyhulme, and later Salford sewage treatment works.

Ardern and Lockett were using filtered compressed air at a rate 0.006 to 0.009 m^3/L (1 to1.5 ft^3 of air/gallon) applied via porous tiles; thus they were using fine bubbles, diffused air aeration. Other methods of aeration were developed along with hybrid versions of this process. Surface aeration by mechanical means became popular and coarse bubble aeration was viewed with some degree of success.

A coarse bubble aeration plant, which was commissioned at Salford in 1914, had an initial capacity of 54.5 m^3 and finally 154.5 m^3. It was operated on a fill and draw basis and could treat 363.68 m^3 per day of a very strong industrial sewage.

As a result of larger scale experiments it was established that the activated sludge process was a most practical and efficient method of treating sewage.

The very first full-scale activated sludge plant was installed at Worcester in 1916 by Jones & Attwood Ltd and treated 3.4 ML per day of sewage which received only screening and grit removal before being applied to the activated sludge unit. Operation was initially by fill and draw method, later changing to continuous flow. The returned sludge was not re-aerated, but received sufficient aeration from being transferred from settlement tanks to the activated sludge unit by air lift pumps (Bolton and Klein, 1971; p. 37).

This new and revolutionary method of sewage treatment discovered by Ardern and Lockett was exploited in many different forms. The advantages of activated sludge treatment when compared with biological filters are listed here.

1. Much reduced land area required.
2. Control of the volume of biomass relative to the flow of incoming sewage.
3. Control of the retention time of the sewage to be treated.
4. Control of the volume of air which could be mixed with the incoming sewage.
5. As a result of control over biomass retention and air supply, control of the degree of treatment to be effected.

SURFACE AERATION

In 1912, J. Haworth was experimenting at Sheffield with the direct aeration of sewage by diffused air. He was, in fact, experimenting with coarse bubble and spray aeration within contact beds. The discovery of the activated sludge process by Ardern and Lockett was a great encouragement to Haworth to continue with his experiments.

From the findings of Ardern and Lockett he was aware that the air performed two main functions in the process. These were the supply of oxygen in a soluble form to the organisms necessary for purification, and also to mix the sewage and circulate it well with the active sludge responsible for the purification. Upon learning that the majority of the air being supplied in the diffused air process was being utilised to circulate and mix the active sludge with incoming sewage, and that very little was being used to maintain the aerobic bacteria responsible for purification, Haworth decided to separate the two functions.

Haworth was a man of some vision and had readily appreciated the natural purification that occurs in rivers. After receiving sewage effluent, a river will flow for several kilometres during which time surface aeration occurs and the river becomes self purified. Haworth's object was to simulate and accelerate these conditions within the confined space of a sewage treatment works.

Haworth's first experiment resulted in the use of a paddle wheel to provide the mixing of activated sludge with the settled sewage to be treated, and achieved aeration by a separate mechanical means. It should be mentioned that Haworth had experienced earlier problems using diffused air due to blockage of the diffusers. It was thought that this was probably due to the nature of the sewage at Sheffield at this time.

Haworth found that the mechanical aerator was unnecessary since sufficient air was being provided by surface aeration and the paddles were providing adequate mixing. This led Haworth to report that 'the results of this and other experiments led to the conclusion that given suitable agitation to obtain thorough intermixing of the activated sludge with the sewage, the necessary air required to maintain biological activity might be obtained by surface aeration' (Institute of Sewage Purification, 1954a; pp. 174–186).

Haworth designed two further experimental plants which were the forerunners of the continuous channel approach which forms the basic 'bio-aeration' plant as his process came to be known. He divided two rectangular tanks by means of partition walls into two systems of continuous flow channels, some 32 m in length by 1 m wide. The motive force for circulation through the channels was provided by two vertical pistons which operated on the same principle as the Roots Blower. One of the channelled tanks was fitted with a mechanical aerator which was found to be unnecessary and this plant operated very successfully for a period of over 2 years on a fill and draw basis of three fillings per day.

Following the success of this experiment a much larger experimental plant was built at Tinsley and commissioned in 1920. This tank was divided into eighteen channels 1.2 m wide and thus the sewage travelled up and down many times throughout the length of the tank. It was ultimately connected by a transverse channel carrying the sewage back to its starting point to continue its route through the system. The continuous channel formed in this way had a length of almost 1.35 km (Institute of Sewage Purification, 1954a; p. 272).

Two shafts were arranged to cross the centre of this aeration tank, carrying 18 paddle wheels, each one was 3 m in diameter. These provided the agitation and circulation power to convey the sewage through the channel system where surface aeration took place. This plant proved to be a most successful single-stage sewage treatment process.

The main sewage works at Sheffield, completed in 1914, had deteriorated somewhat over this period, mainly due to shortages during the war years (the 1914–1918 war), such that extensions and new works were desperately needed. As a direct result of the success of the Tinsley plant it was decided in 1921 to build a large 'bio-aeration' unit to treat Manchester sewage.

The new plant was built on a design very similar to the plant at Tinsley. There were a few engineering differences. The channel width was increased to nearly 2 m and the total channel length increased to 1.8 km for each aeration unit. Since Haworth had proved that the exchange of oxygen at the interface of a liquid with the atmosphere was adequate to maintain biological activity, the artificial river concept was supported.

In 1954 Sheffield had some 45 km of channel treating a dry weather flow of 109 ML per day. The process was known as the 'Sheffield System' as well as the 'Haworth Paddle System' and 'Bio-aeration' (Institute of Sewage Purification, 1954a; p. 272).

SIMPLEX CONE AERATION

The Simplex aeration process was developed at Bury Sewage Works by Joshua Bolton. Bolton was aware, due to contact with other workers in the sewage treatment field, that direct aeration of crude sewage or settled sewage had beneficial effects. In most cases these experiments were only successful after long periods of aeration and it was not until the revolutionary work of Ardern and Locket that retention of the active sludge was realised to be essential to any practical purification process.

From November 1914 to May 1915 Joshua Bolton conducted experiments at Bury with activated sludge at Bury sewage works using direct aeration and achieved between 80 and 90% purification, in terms of BOD. It must be emphasised these experiments were laboratory scale, involving 4 L jars aerated with a perforated coil.

Bolton then conducted experiments on a larger scale utilising the existing tanks at Bury sewage works, which were 6.0 m deep and 8.0 m in diameter and had with a total capacity of 227 m^3. Initially aeration was by compressed air forced through brass nozzles at the base of the tanks. He experienced much trouble with blockage within the pipes. Whenever the air supply was interrupted, sludge with a pressure of 6.0 m of head was forced into the aeration tubes through the nozzles. The distribution of air throughout the tank was also very uneven. Bolton continued to experiment with diffuser tiles and had similar problems to other workers in this field of blinding and eventual blockage of the diffusers.

At a meeting of the Association of Sewage Works Managers' in July 1916, Joshua Bolton was impressed with the work being undertaken by John Haworth, using revolving paddles. In November 1916 Bolton experimented with a water wheel propelling sewage and activated sludge along shallow channels in a 4 L laboratory scale system. This was operated on a fill and draw basis and the results were encouraging. An average purification of 88.8% was achieved. The depth of liquid in the channels was 5.0 cm and this meant that a high surface area was exposed to the atmosphere, far more than could be achieved in practice on a full-scale plant (Institute of Sewage Purification, 1930).

The only tanks available for adaptation and experiment on a large scale at Bury were six circular tanks, with dimensions of 8.0 m diameter and 2.6 m deep. Before attempting to utilise these, Bolton constructed a tank of approximately 1.0 m^3 capacity from an old boiler shell with a conical base (see Figure 5.3). This had a central uptake tube flared at the base and a circular inverted cone at the top. By placing an air line down the central tube an air-lift was created which drew the contents to the centre, up the central tube and over the circular weir near to the surface of the tank. The thin film of liquid passing over this central column and the resultant splashing as this fell to the surface produced good aeration and was a most successful experiment.

Encouraged by this Joshua Bolton constructed a concentric inner shell within this unit to provide an annular settlement area. The settled sludge gravitated to the central circular tank for agitation and aeration with incoming sewage. This second experiment was a continuous flow system. As sewage entered it displaced clarified liquor over the annular weir surrounding the tank. This system achieved 95% purification.

Bolton was a man of great ingenuity with a good grasp of the fundamental principles of this process. He realised that to conduct a full-scale treatment plant by this method would not be practical as many vertical pipes would need to be installed within his large circular tanks and a single central tube would be too large in diameter for an air-lift to operate. He approached this problem by fitting a single large uptake tube in the centre of one of his 8.0 m diameter tanks and raised the flow through this tube by a scoop wheel. This was a mechanically operated wheel much like a very small, circular, bucket dredging device.

This method proved to be unsuccessful since, although it circulated the activated sludge throughout the tank, it provided little or no aeration. Even a later model, utilising baffles strategically placed near the tank's central surface was unsuccessful. He found that by fitting a revolving cone with external vanes on top of the uptake tube, liquid was thrown off the periphery of the cone producing good aeration. Liquor was easily drawn up the central tube by this method, thus affording good circulation in addition to efficient aeration (Institute of Sewage Purification, 1930).

This system proved so effective that it gained much popularity and full-scale systems were installed at Burnley and Birmingham, as well as Bury. The first full-scale cone aeration plant was purpose built at Bury and commissioned in 1925. It was designed to treat a dry weather flow of 3637 m^3 per day and consisted of three units each containing eight cones operating in series. In the same year a plant was installed at Epsom to treat flow of 2273 m^3 of settled sewage per day.

The design of the cone itself, especially the angle of the cone and the configuration of vanes upon its outer surface was improved considerably over a period of some 20 years and a 'high intensity' cone was introduced in 1946. The effect of this new design enabled the capacity of a plant to be increased by some 50%, producing the same quality of effluent in 12 hours that was formerly achieved in 18 hours with the same power consumption per unit quantity treated. Many existing simplex cone aeration plants were improved by installation of new high intensity cones (Institute of Sewage Purification, 1954b; pp. 48–63).

THE KESSENER BRUSH AERATION SYSTEM

Dr H. Kessener was Director of the Government Institute for Waste Water in The Hague, Holland. He conducted a series of experiments in 1925, trying to achieve surface aeration by means of brushes. Originally these plants were meant for intensive purification of industrial wastes from isolated factory units in Holland.

A shaft fitted with brushes was suspended above the water surface in a rectangular tank in which sewage and activated sludge was circulated. Aeration was by means of rapid rotation of the shaft and hence the brushes through the surface of the liquid. The horizontal drive shaft with its configuration of brushes was suspended along one side of the aeration tank such that the tips of the brushes were immersed to a depth of some 50 to 100 mm. Circulation was aided by the cross-sectional design of the tank and the judicious location of baffle boards. This configuration imparted a spiral forward motion to the contents of the tank and the mixed liquors moved from one end of the tank to the other. Rotation speeds varied from 50 to 100 revolutions per minute. The depth of immersion of the brushes in the liquid was achieved by movement of the shaft or control of water level within the tank.

Kessener's first experiments were done with Piasava bristle rotors, but these tended to snap at high rotation speeds and were replaced with stainless steel combs. The diameter of the Kessener brush was approximately 50 cm and depth of immersion was variable, but was usually within several centimetres of the water surface.

In 1926 Kessener designed such a plant for an installation at Apeldoorn to treat 45 m^3 per day of abattoir effluent, a very strong trade waste. This was the first successful, commercial application of this system (Bolton, 1954).

Dr Kessener patented his Kessener brush system of aeration in Britain in May 1935. The first installation of such a plant in the United Kingdom was commissioned at Stockport in 1937. This plant consisted of two aeration units with stainless steel combs rotating at 140 revolutions per minute. The brushes were immersed to a depth of 75 mm. The first aeration unit was for treatment and the second for re-aeration of the sludge for re-use in the first unit (Stanbridge, 1976; vol 7, p. 46). Other installations followed in Rochdale and much later in Manchester in 1955. Over a period of time many changes were made in the design of the rotor and cross-sectional shape of the aeration unit. (Kessener 1930).

Many methods of surface aeration were the subject of experiments over the period from 1918 to the late 1920s. Only those systems mentioned above were regarded as being successful and the development of cones, paddle wheels and rotors continued. These are the only forms of mechanical aeration which are generally still in use today. The popularity of the paddle aeration system tended to decline due to the large land area required for this in relationship to the smaller areas necessary for cone and rotor systems.

With the exception of Sheffield no paddle plants were installed after 1935, but the installations at Sheffield and Stockport were still operating successfully in 1972. The Blackburn Meadows plant has only recently been dismantled in 1993. Some paddle bio-aeration plants were unsuccessful.

It is interesting, especially when considering the work of Mumford and the iron coagulating organism 'M7', that the Sheffield plant always produced quite good results. This could be attributed to the characteristic presence of iron in Sheffield sewage from the extensive local steel industry, since iron salts are known to be good flocculating agents.

Chapter 7
Later developments

BACTERIA BEDS

As a result of the work carried out by Corbett, and later Candy and Caink, and the widespread adoption of bacteria bed treatment, modifications and further improvements have been made to this process.

Despite the fact that the introduction of bacteria beds meant that a smaller land area was required to treat the same volume of waste than that formerly needed for land treatment, there was a constant need to reduce the land area required to treat sewage from the larger populations. Increasingly this need was accompanied by a requirement for better quality effluent and hence more efficient treatment of the sewage. Available building land near to large communities was expensive and with an increased population, and hence higher volume sewage flow, the impact on a receiving watercourse was greater and therefore treatment to a higher degree was demanded.

ALTERNATING DOUBLE FILTRATION

This led to several operational variations of bacteria bed treatment. One of these, originally developed for treating strong dairy waste, was that of alternating double filtration (ADF). This process involves passing settled sewage through one or more bacteria beds, then achieving humus settlement by passage through a settlement tank and returning the flow to a second bed or group of bacteria beds.

After several days the initial bed builds up a high level of biomass and will eventually begin to clog and 'pond' and it is then necessary to reverse the procedure and the bed with the higher build-up of biological film becomes the secondary bed. The biological film accumulates and then disintegrates by alternating the primary and secondary bacteria beds, and there is no undue build-up of film. This method was applied to domestic sewage treatment at Birmingham at its Minworth sewage works over a period of three years, from 1938 to 1941. The results of this application of alternating double filtration are detailed in a paper by Wishart and Wilkinson (Stanbridge, 1976; p. 55).

It was shown that at least double the volume of settled sewage could be efficiently treated in this way, thus the capacity of a sewage plant could be doubled without any need for additional bacteria beds. This was a most dramatic step forward even after allowing for the construction of extra humus tanks and taking into account the return pumping costs involved. The costs of doubling the area of bacteria beds was much higher than the smaller capital cost involved by providing ADF and this system could be easily incorporated into existing sewage works.

This adaptation of bacteria bed treatment became very popular and was adopted at many other sewage treatment works, especially those which were overloaded and needed to quickly and cheaply be modified to treat larger or stronger volumes of incoming sewage.

HIGH RATE FILTRATION

In many cases the modification and improvement to the bacteria bed process was instigated by the need to treat wastes of high organic strength which often resulted from the trade effluent content of sewage from municipal drainage areas. This was the case with high rate filtration, whereby a bacteria bed was dosed at up to 30 times the normal standard dosage rate. In other words, sewage could be applied to such a filter at rates of up to 12 m^3 per m^3 of media per day (Kessener, 1930). Even higher rates of application could be achieved by the re-circulation of treated effluent to mix with the incoming sewage and thus dilute its strength.

A high-rate filter would be incorporated in a sewage treatment process for the rapid removal of BOD (some 50% of BOD removal could be achieved in one passage through such a filter). The partially treated effluent would then be applied to further bacteria beds to complete the treatment process.

Rarely were high-rate percolating filters used to achieve nitrification. Such bacteria beds were certainly a useful and novel application of the basic principle. The cost of bed construction was much reduced by the introduction of synthetic plastic media. Hitherto bacteria bed media was usually of a mineral nature, being granite chips or similar material. The use of manufactured, lighter plastic media of varying designs, which proliferated from the late 1950s, enabled weaker supporting structures.

This led to the evolution of high-rate biological towers since the weight of such media was nowhere near as heavy as that of conventional mineral media. The walls and retaining structures of a bacteria bed could be much thinner and lighter and therefore cheaper. It was possible to construct very high bio-towers which therefore had good ventilation characteristics and also enabled extensive zoning within the bacteria bed for carbonaceous oxidation and full nitrification.

COVERED BACTERIA BEDS

It was also discovered to be quite advantageous to enclose a bacteria bed by constructing a roof which was air tight. Such beds would have a depth of some 4 to 6 m. It was necessary to provide a forced draught in order to ventilate the bed which was done by introducing a fan and drawing air downwards through the media of the bed.

This type of bacteria bed had many advantages. There was less nuisance from flies or odours since neither could easily escape from this system and since birds could not prey on the bed surface, the scouring organisms in the top few centimetres of the bed could happily exist all the year round. The bed was less subject to the normal seasonal fluctuations experienced in conventional bacteria beds. The presence of scouring organisms reduced the possibility of clogging the bed.

Several plants utilising these modified bacteria beds were constructed in Germany and also in South Africa. At the Rondebult sewage works in Germiston, South Africa, over a 3-year period, the enclosed aerated filter was found to be capable to treating almost two and a half times the volume of sewage to the same degree of quality as that of an open filter of some 2 m in depth. At this works ADF was also introduced by using two enclosed filters in series and reversing the order of the primary and secondary filter on a regular basis. As a result the treatment capacity was increased to almost three and a half times that of a conventional bacteria bed (Stanbridge, 1976; vol 7, p. 66).

Further experiments were carried out with enclosed aerated bacteria beds at Dalmarnock sewage works in Glasgow, Scotland, over a 4-year period by Hunter and Cockburn (Wishart and Wilkinson, 1941). Their work was probably more reliable since it was carried out over a longer period and under more controlled conditions than the work done in South Africa. It was concluded that twice the volume of sewage per unit volume of media could be treated by the enclosed bacteria bed when compared with a normal bacteria bed producing effluent of the same quality.

PACKAGE PLANTS

In the late 1970s the combination of plastic media and the covered bacteria bed was employed for many types of smaller 'packaged' sewage treatment plants for treating the waste from small communities and single dwellings.

Such enclosed bacteria beds with plastic media, supplied with air by a small electric fan were often housed in multiple-sectioned, glass-reinforced plastic (GRP), package treatment units. These were supplied with small electric pumps for irrigation of the plastic media and return of settled humus sludge to the primary settlement zone.

Thus, a complete sewage treatment unit could be encapsulated in a very small volume and even be installed unobtrusively below ground level. Such plants are presently available from several manufacturers for use at caravan sites, public houses, army camps, hotels and private houses.

ROTATING BIOLOGICAL CONTACTOR

The rotating biological contactor (RBC) process has been extensively developed from the first experiments undertaken in America by J Doman in 1925.

Multi-disc arrangements on a single shaft in a covered GRP unit have become very popular. Manufacturers have experimented with different configurations of disc surface (spiral, corrugated, etc.) to increase the surface area of the disc and also different rotation speed and depths of immersion have been tried. These plants are very widely used as packaged treatment plants for the smaller community and have also enjoyed much use as tertiary treatment for smaller sewage treatment works.

Modern RBCs can achieve very high quality effluents with relatively low retention times and also have a small space requirement ('foot print'). Much has been achieved by the use of modern materials, such as glass reinforced plastic, high density polythene, and plastics generally, to fabricate quite large areas of disc which do not exert high stresses on the shaft since they are light in weight. These materials do not corrode and require little maintenance.

ACTIVATED SLUDGE

The activated sludge process is now available in a wide variety of forms. Not only has the application of fine bubble and coarse bubble aeration been exploited in many ways, there have also been several variants of mechanical aeration. Improvements have been made over the years to the Kessener brush and simplex cone surface aeration systems. By optimising the design of aeration tanks for the differing aeration principles, treatment efficiencies and retention times (and hence plant size) have been much improved.

PARTIAL ACTIVATED SLUDGE TREATMENT

One of the most common applications of the activated sludge process, especially initially, was its introduction into sewage treatment as a partial process. It had a small 'footprint' and could be used to extend the capacity of existing sewage treatment plants. Activated sludge had many advantages in that it would rapidly cope with the colloidal matter present in incoming sewage and also carry out the carbonaceous oxidation most efficiently.

The second stage of treatment, usually upon percolating filters (bacteria beds), led to a well nitrified effluent since these bacteria beds would be capable of nitrification in their upper zones, usually reserved for carbonaceous oxidation.

Many municipal sewage treatment plants incorporate activated sludge as a partial stage with great success.

AERATION

Different techniques of operation have also been introduced for the activated sludge process. Modified forms of aeration have been used to provide better utilisation of air for reasons of capital economy, as well as to provide more flexible control and reduce operational costs. The supplying of air to activated sludge, whether by a diffused air process or mechanical aeration, is the predominant operating cost of the activated sludge process. Oxygen requirements in this process are greatest at the inlet of aeration tanks where the incoming sewage enters the plant, and the requirements are least at the outlet. Tapered aeration has been practised by providing increased air in the initial stages of the process and reducing the air supply in the final stages. Effective treatment can be carried out and air supply costs can be much reduced (Tebbutt, 1983; pp. 151).

HIGH RATE ACTIVATED SLUDGE TREATMENT

As with bacteria beds it has also been found that activated sludge plants can be made to work on the ‘high-rate’ principle. A ‘high-rate’ activated sludge process involves the use of much less active sludge in the mixed liquor and quite short aeration periods. Instead of the more usual 2000 to 3500 mg/L mixed liquor suspended solids, such plants will operate on very low mixed liquor suspended solids levels of around 600 mg/L. Retention times in ‘high-rate’ activated sludge processes are usually less than 2 hours and the effluent produced can achieve a reduction in BOD of some 75%.

The loading of such plants is usually about three times that of conventional plants. Such systems are most useful when there is little space available. These systems can rapidly reduce the load of incoming wastes for subsequent treatment on bacterial beds, conventional activated sludge plants or tertiary treatment.

The ‘high-rate’ activated sludge process has other advantages in that the activated sludge that develops in such systems is quite dense and settles readily and for this reason does not often cause troublesome ‘bulking’ which may sometimes occur in conventional plants (Dekema and Murray, 1942).

EXTENDED AERATION

Extended aeration, sometimes known as ‘total oxidation’ or ‘aerobic digestion’, is yet another variant of process operation. By extending the aeration period (or increasing the retention time) to be 24 hours or greater, and introducing a high rate of return sludge, considerable oxidation of the activated sludge solids takes place. The rate of accumulation of sludge is much smaller than for conventional operation and this has the advantage of less surplus sludge being produced for subsequent disposal. Because of this, the extended aeration plant is somewhat simpler to operate and has been used to great advantage in smaller self-contained packaged sewage treatment units. Such systems need to operate with much higher suspended solids levels in the mixed liquor and plant loadings are usually much lower than those of conventional activated sludge systems.

This system produces a well nitrified effluent and has been stream-lined in its design for application to small communities and provides good effluent quality and minimal maintenance.

The activated sludge system, although very efficient, needs to be controlled in such a variety of ways (sludge concentration, retention time, degree of aeration) that few plants have been developed for the smaller packaged market, where they must be simple to operate by sometimes, non-technical personnel. Extended aeration reduces the need for regular sludge removal, and this can be done at service interval frequencies, when such a system is applied to packaged units. There have been two other major innovations to conventional activated sludge processes and these are the deep-shaft technique and direct oxygen injection.

DEEP SHAFT TREATMENT

The deep shaft technique was introduced in the 1970s. It is a high intensity aerobic treatment technology, ideal for reducing the BOD load of very high strength sewage and industrial wastes. The technique lends itself to introduction within existing sewage treatment facilities too.

The treatment process comprises a single deep vertical shaft which can be up to 6 m in diameter and between 40 and 100 m deep. Surrounding the top of this shaft is the de-gassing tank. The plant could best be described as a vertical activated sludge plant in which the flow passes down one side of a deep hole and returns up the other side of this shaft. The incoming flow is mixed with activated sludge and is driven by the injection of compressed air down one side of the shaft. It then passes up the opposite side of the shaft and takes between 2 and 5 minutes to circulate once round this system.

Typical retention times are between 20 and 40 minutes. The flow circulates rapidly in the shaft since it is driven by the differential density of liquor in the riser and the down-comer of the shaft. Because of this a very turbulent flow results and this provides intense mixing of air, liquid and biomass.

In such treatment plants primary sedimentation is not necessary and incoming sewage can be fed to a deep-shaft activated sludge plant after basic screening. Because most of the process takes place in a vertical shaft in the ground, this plant has a very small land area requirement. Since this system needs no primary settlement this also adds to the attractive economic advantages of deep-shaft technology.

The process is capable of treating a wide range of liquor strengths and can operate at very high mixed liquor suspended solid concentrations. The system also has a high oxygenation efficiency of up to 60%. The reason for this is that during the 2-minute resident time in the shaft the mixed liquor and the process air bubbles remain in contact for approximately one and a half minutes. This is a long residence time when compared with conventional diffused air plants where bubble retention time is only several seconds.

Due to the highly efficient aeration, the mixed liquor is supersaturated with nitrogen, carbon dioxide and some residual oxygen and micro-bubbles will also be attached to the active sludge flocs. These need to be removed prior to separation of the activated sludge and the treated liquor since their presence will impede settlement. This can be achieved by use of a de-gassing tower in which the mixed liquor is subjected to reduced pressure. Another method is to allow the flow to pass to separate tanks where the mixed liquor is subjected to very fine bubble aeration which rapidly causes the entrained micro-bubbles to float free from the sludge flocs.

Separation of the mixed liquor suspended solids and the treated effluent then takes place by gravity in conventional clarifiers or settlement tanks. Due to its low residence time, low aeration tank volume and high BOD removal rates, this type of treatment has many advantages, not least of these is the overall cost effective high performance (Hunter and Cockburn, 1944; Bolton and Klein, 1971; p. 96).

OXYGEN SYSTEMS

The use of oxygen instead of air in activated sludge systems was introduced in the 1970s. This intense oxygenation had several advantages. It enabled very high strength waste to be treated and also, to a certain extent, protected activated sludge systems from shock loads, especially those emanating from industrial wastes. There were other benefits which were also found to result from direct oxygen aeration, such as the production of less surplus sludge and also the active sludge had much better settling characteristics, even at quite high loadings.

The ability to provide adequate concentrations of dissolved oxygen and thus cover a wide range of variations in influent loads enabled these systems to produce superior quality effluents and enjoy considerable stability of operation while doing so. Initially, problems were encountered with the production of carbon dioxide from the carbonaceous oxidation phase of the process and this, in many cases, remained in solution and thus lowered the pH of the mixed liquor. At certain low pH levels this was inhibiting to nitrification. This problem was encountered mainly, however, with early covered tank systems.

Apart from introducing oxygen directly through the aeration system, another technique was developed whereby oxygen-saturated water was introduced into the activated sludge plant. This particular variation of the process is known as 'side-stream' aeration. Oxygen-activated sludge systems are more widely used today, but more specifically for troublesome wastes, especially those industrial wastes which are pre-treated at source prior to disposal to the foul sewer or to a watercourse (Tebbutt, 1983; p. 156). Oxygen systems have applications for the treatment of strong and difficult wastes such as those from the paper pulp, pharmaceutical, and food production industries.

COMBINATIVE TREATMENT METHODS

A quite fascinating concept, which has seen wide acceptance in the field of smaller packaged sewage treatment plants in the last ten years, is that of fixed film activated sludge treatment.

Such systems are sometime known as submerged-aerated filters (SAF) or submerged-aerated media or biological-aerated filters (BAF), but they are essentially activated sludge plants provided with a matrix upon which the activated sludge can cling and develop. Such plants are an interesting combination of the activated sludge process with the percolating filter or, in some cases, RBC systems.

SUBMERGED AERATED FILTERS

There are currently several manufacturers producing SAF units for the packaged sewage treatment market. These plants are ideal for treating wastes from small communities, hotels, public houses, etc. Many BAF plants have been used successfully for treating high volume and high-strength trade effluents (Figures 7.1 and 7.2).

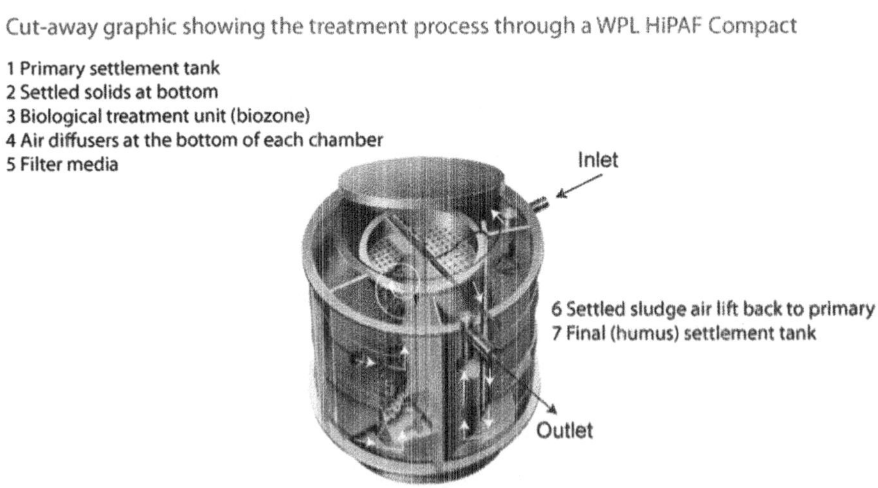

Figure 7.1 A modern compact submerged aeration plant (Courtesy of WPL Ltd)

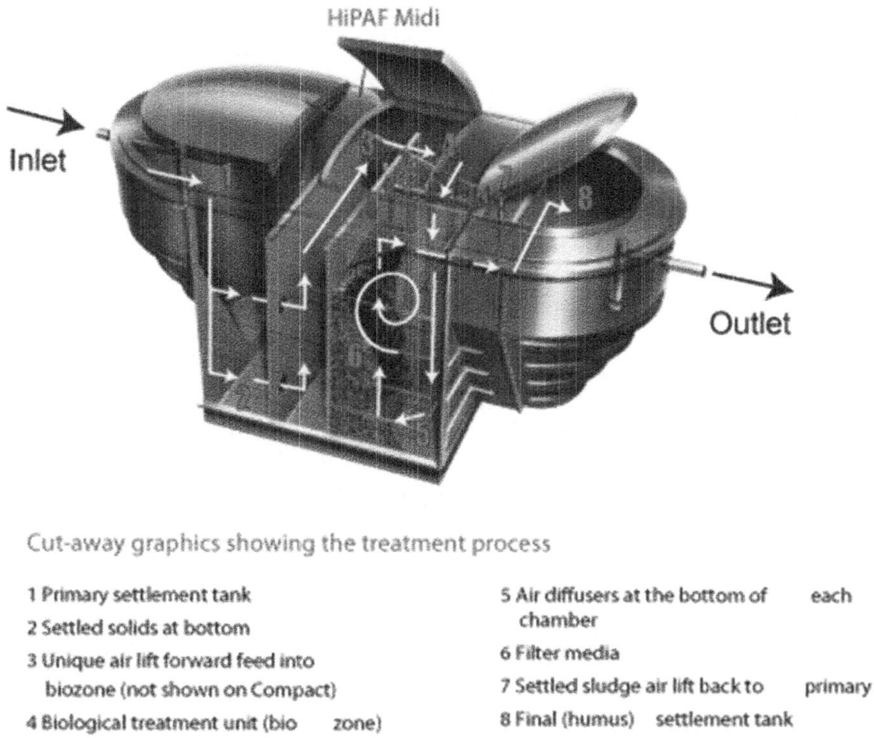

Figure 7.2 A larger submerged aeration system (Courtesy of WPL Ltd)

The basic treatment is usually fine bubble aeration which is introduced below high voidage plastic filter media which is submerged in settled sewage. Retention times are not uncommonly high and a biomass quickly develops both on the plastic media surface and also as a free-floating 'true' activated sludge floc. The flow for treatment can be introduced at the base of the system for upward flow or a down-flow system can be employed. Generally BAFs have a back wash cycle to scour excess biomass from the media and so prevent blockage. This allows high loadings to be applied (4 to 8 kg of BOD per cubic metre per day) and this combined with their small land area requirement (footprint) has led to their use for trade effluent treatment. SAF plants normally operate without an air scour facility and are simpler, low maintenance, more robust systems, and are ideal for the package plant market. SAF loadings are commonly 0.5 kg of BOD per cubic metre per day.

These plants have become very successful and many have quite minimal maintenance and few moving parts. Some are powered by just a single blower unit which provides air for the treatment zone as well as supplying pumping by means of an airlift to re-circulate effluent or return settled activated sludge to the primary settlement tank. More sophisticated units sometimes have integral sand filters and employ air from the same air source, controlled by electronic timers, to provide an air-scour back- wash sequence.

A unit of this type, with built in tertiary treatment from such a sand filter can produce an effluent of <5 mg/L BOD <5 mg/L suspended solids and <10 mg/L ammonia. These plants are really a combination of a fixed film aerobic bacteria system (percolating filter) and a suspended growth or dispersed floc system (activated sludge). The method combines the stability of fixed film treatment with the high transfer rate and purification potential of the dispersed system.

Even these systems can be improved by encouraging a zoning effect and separating the high voidage plastic media, which is submerged and aerated, into two or more zones, such that the flow passes from one to the other and finally to a solids settlement area. In this way colloidal breakdown and carbonaceous oxidation can be encouraged in the first bio-zone, while the second aeration phase can effectively oxidise ammonia fully to nitrate.

Due to independent development by different manufacturers there are many variations of this process.

AERATED RBC

The other major fixed film biological reactor, the rotating biological contactor (RBC), or the rotating disc system has also been adapted in a similar way to operate in conjunction with an independent air supply introduced below the surface of the mixed liquor.

Initially diffused air was introduced into the bio-zone of RBC units which were operated in the conventional way, having some 40% disc submergence with supplemental air supplied by diffusers. This was further improved by the construction of plants which had between 80 and 90% disc submergence which were also provided with a diffused air system.

A later adaptation was the utilisation of the rising air to drive the discs and thus remove the need for external motors and gear boxes. This was achieved by fitting the discs with peripheral air cups to capture the rising air from carefully positioned diffusers at the base of the tank. Originally this system was used with RBCs having 40% disc submergence, but it was found to be much more beneficial to almost completely submerge the discs and thus provide a rotating fixed film-activated sludge system (Collins and Elder, 1977).

Due to the buoyancy of a shaft which had hundreds of discs each having air cups, the structural loads were reduced by up to 80% with consequent gains in mechanical efficiency. It was also found that the air scour across the surface of the disc tended to control microbial growth and prevent too thick a biological film developing on the disc surface. This enabled more discs to be packed on a single shaft and thus vastly increased the available surface area of a single unit.

A large submerged biological contactor plant was installed by North West Water at Partington in Greater Manchester in 1992, and it has reportedly been most successful (Ousby JC et al., 1977).

CONSTRUCTED WET LANDS

A most novel aerobic treatment which has been introduced in the UK and Europe over the last decade is that of the Constructed Wet Land, Root Zone or Reed Bed treatment. Original work on this system is claimed to have been carried out by Professor R. Kickuth of Kassel University, Germany. The system was also developed independently by the Water Research Centre at Medenham. It is now widely used for tertiary treatment of treated sewage effluents and also, in many cases, for the treatment of settled sewage and various industrial wastes. Kickuth developed formulae for the area of reed-bed required for differing volumes and strengths of waste and also formulae to calculate the optimum cross-sectional flow and optimum bed widths to give the best flow characteristics through the reed-bed.

Quite large land areas are required for this type of treatment. The reeds most commonly used are *Phragmites australis*, which are native to the United Kingdom. These are planted in gravel beds and the waste to be treated is passed through the gravel. The waste flow is below the gravel surface and aeration is carried out by the rhizome systems of the reeds and the tubular structure of the reeds themselves.

The most usual method of application is the horizontal flow reed bed but vertical flow reed beds are also used. The inlet to a vertical reed bed is much higher than the outlet, hence the vertical flow. The horizontal flow bed has an outlet only marginally lower than the inlet level. There is usually provision for flooding the bed to a depth of some 400 mm in order to control weed growth between the reeds. This is a seasonal operation and considerably reduces maintenance of the reed-bed (Boon, 1976).

Such systems have become very popular in recent years for both small installations and larger ones. Severn Trent Water Ltd have adopted a policy of RBC and reed-bed treatment for many of their small, unmanned, rural sewage works. Examples of much larger installations would be those undertaken by ICI at Billingham where a 5 hectare reed-bed has been installed. British Steel operate a system at Newport in Wales which occupies an area of 18 hectares. This latter is intended to treat phenolic effluent from their steel works.

Great claims are made for these systems, but unfortunately a reed-bed needs between 3 and 5 years to achieve its full potential since the reeds need to reach maturity. Although running costs and maintenance of these reed-bed systems can be very low, capital costs can be surprisingly high for the preparation, planting of the reeds and commissioning.

The system is really an extension of early land treatment of sewage, but is more controlled. Rhizome action and transpiration of the reeds provides aerobic conditions for the development of organisms which treat the sewage as it passes through the gravel bed just beneath the surface (Chi-Su Chou and Hynek, 1979).

Reed-beds and wetland treatments have seen resurgence but due to the required land areas these have limited applications. These systems are slow to acclimatise to fluctuations in load and volume. They can also be susceptible to inhibition by toxins. A reed-bed site could be the source of future problems since low-level, persistent toxins ‘removed’ by these systems will accumulate and concentrate within the root zones.

Chapter 8

Influences upon sewage treatment

INDUSTRIAL WASTES

The presence of iron salts from trade effluents in the sewage of Sheffield may have aided the formation of sludge flocs and thus had a beneficial effect on activated sludge treatment processes. Unfortunately, in the majority of cases, discharges from industry to sewage treatment works were detrimental. Trade effluent discharges, especially before the instigation of controls, and before the widespread use of effluent pre-treatment systems, were very damaging to biological sewage treatment processes.

Around the 1850s the rapid development of the industrial revolution resulted in numerous trade effluents discharging to watercourses. The cotton industries of Lancashire as well as tanneries, paper mills, chemical works and later gas works, produced extremely polluting wastes. Many industries had become established on the banks of rivers to take advantage of the motive power for water mills and the use of process water. It was therefore convenient to discharge liquid wastes from industry directly to the river.

EARLY LEGISLATION

The Rivers Pollution Prevention Act 1876 formed the basis of all legal action connected with the pollution of rivers until 1951 when the Rivers (Prevention of Pollution) Act was passed. The Act of 1876 was the first in which manufacturing industries were permitted to discharge their trade wastes to a local authority sewerage system (part IV Rivers Pollution Prevention Act 1876) (Forster, 1985). The only restrictions to discharge trade wastes to a public foul sewer under this Act were that the wastes must not adversely affect the sewers and also that the sewerage system must be large enough to accommodate the waste flow. Apparently the onus was on *conveying* the waste, not necessarily *treating* it.

The Royal Commission on Sewage Disposal, which was originally appointed in 1898, addressed the problem of trade effluents in its third report of 1903. In this report the Commission stated 'the law should be altered so as to make it the duty of the Local Authority to provide such sewers as are necessary to carry trade effluents as well as domestic sewage, and that the manufacturer should be given the right, subject to the observance of certain safeguards, to discharge trade effluents into the sewers of the Local Authority if he wishes to do so'. The report went on to add 'the Local Authority should frame regulations which should be subject to confirmation by a central authority and could provide definite standards for the different manufacturers as regards preliminary treatment' (Cooper *et al.*, 1989).

This is the first reference to any requirement for industrial discharges to be controlled prior to discharge to the public foul sewerage system.

CHARACTERISTICS OF TRADE WASTES

Most large towns had trade effluent discharges and the characteristics of these reflected the local trades and industries. There were, however, certain common trade effluents which would be present in the sewage discharge from most reasonably sized communities. The more universal wastes would be dairy waste from the bottling of milk to serve the local community, abattoir waste where a town had an abattoir to supply fresh butchered meat to the locality and most towns of any sizeable population would have their own gas works to supply gas for domestic and industrial use. Gas works produced liquid wastes containing particularly high levels of phenols and ammonia.

Other trade effluents predominated in the densely populated and highly industrialised areas of England, such as Yorkshire, the Midlands and Lancashire. The cotton trade of Lancashire produced many different trade effluents such as bleaching, dyeing, textile printing, textile sizing and laundering. The woollen trade in Yorkshire gave rise to greasy, dirty waste from wool washing and associated fell-mongering and tanning wastes.

Examples of wastes discharged in towns and cities throughout the country would include those from tar distilling, paper making, brewing, food processing, electro-plating and chemical manufacture. The range of different wastes which could inhibit or be toxic to sewage treatment was therefore quite extensive.

TRADE EFFLUENT STRENGTHS

The resultant load, additional to domestic sewage, exerted on biological sewage treatment units by these different trade effluents is better appreciated by reference to Table 8.1.

The 'load' is expressed in terms of population. In order to measure the anticipated daily loadings from such industries the population equivalent is defined, this figure is related to a throughput of product (e.g. 1 ton of manufactured yarn, or 2 cm^3 of beer).

Table 8.1 Population equivalents typical for particular industries

Class of industry	Unit basis	Population equivalent per unit
Abattoirs	1 cow = 2.5 pigs = 2.5 sheep	20–100
Bleaching house	100 lb (45 kg) bleached textile	6–16
Brewery	100 gal (455 L) beer	100–500
Cellulose manufacture	1 ton (1016 kg) cellulose	about 500
Dairy with cheese manufacture	1000 gal (4550 L) milk	400–1000
Dye house (including sulfur dyeing)	100 lb (45 kg) dyed textile	about 100
Distillery	1 ton (48 bushels, 1000 kg) cereals	about 1500
Flax retting and milling	1 ton (1000 kg) flax fibre	about 1000
Laundry	100 lb (45 kg) washed textile	20–100
Paper milling	1 ton (1000 kg) paper	100–300
Beet-sugar manufacture	1 ton (1000 kg) beet	100–300
Sulfite pulp milling	1 ton (1000 kg) pulp	1000–4000
Tanning	1 ton (1000 kg) raw hide	1000–4000
Wool scouring	100 lb (45 kg) raw wool	70–140

Source: Imhoff *et al.* (1971).

The population equivalent factor is calculated by comparing the chemical strength of a trade effluent with that of domestic sewage, usually in terms of BOD, thus a ratio can be established and a population equivalent calculated. From Table 8.1 it will be seen that a dye house with a throughput of 450 kg textiles per day produces wastes during that day equivalent to 1000 people. Similarly, a tan house tanning 2000 kg of raw hide per day will produce a waste loading equivalent to 8000 people (Parr, 1989).

Activated sludge treatment units and percolating filters could be overloaded by discharges of high strength industrial waste. Treatment was made more difficult by the intermittent nature and high volume of many of these industrial discharges.

TOXICITY OF WASTES

Apart from the increased demands placed upon biological systems by these wastes, there was also another aspect to consider. Many of the trade wastes, both early discharges and those made later in the 1940s and 1950s from the rapidly developing bicycle and automobile industries contained toxic materials.

The most common example was the wastes produced from the electro-plating process. This process was widely used for corrosion protection and ornament of automobiles. The wastes from electroplating included not only strong acids and alkalis, but also cyanides and heavy metals. The metals copper, lead, chromium, nickel, cadmium and zinc were all poisonous to the aerobic organisms upon which sewage treatment depended.

Phenols resulting from the manufacture of coal gas, another toxic compound, have already been mentioned.

Industrial wastes arriving at a sewage works would often be much diluted by the domestic waste from the inhabitants of the drainage area. Unfortunately, very minute amounts of metals, cyanides, phenols and other toxic compounds could adversely affect aerobic sewage treatment organisms. In instances where this occurred there was a consequential drastic reduction in purification from such sewage treatment units. Gas liquor (or coke oven liquor) analysis is shown in Table 8.2. It will be seen that in addition to phenols, there are also cyanides and thiocyanates present in this waste and a considerable load is produced by the very high ammonia levels (Parr, 1989).

Table 8.2 Analyses of some coke-oven liquors

Liquor	Analysis of liquor (mg/L)				
	Free NH_4^+	Fixed NH_4^+	Total NH_4^+	Phenols	CNS^-
A	162	2304	2466	949	488
B	100	333	433	910	160
C	100	3230	3330	840	400
D	90	306	396	592	310
E	340	3110	3450	1470	1380
F	342	1008	1350	2231	580
G	765	1406	2171	2800	252

Liquor	Analysis of liquor (mg/L)			
	$S_2O_3^-$	CN^-	Total Cl^-	Permanganate value (4 h)
A	308	nil	3532	2680
B	154	nil	450	2260
C	175	nil	6410	2600
D	168	26	888	1040
E	262	nil	5890	4800
F	728	nil	2041	5000
G	560	nil	1935	4680

Source: Callely *et al.* (1977).

One milligram per litre (mg/L) of chromium can adversely affect biological treatment in either an activated sludge or a bacteria bed system. One milligram per litre of cyanide can have a similar effect. It was found that by incremental dosing of biological systems with organic toxins, these can eventually be tolerated and broken down by the organisms present. By increasing the dosage, quite high values of such poisonous materials can be rendered harmless by aerobic biological sewage treatment processes. A good example of this would be the high levels of phenols from coal gas production which originally caused great problems at most municipal sewage works. In admixture with sewage and after sufficient acclimatisation, aerobic processes could treat phenolic waste with little or no adverse effects. Aerobic biological systems were also capable of successfully treating cyanides in the same manner.

SYNTHETIC DETERGENTS

This ability of aerobic biological systems to develop different strains of organisms to cope with predominant feedstock characteristics is a most considerable advantage of this mode of treatment. Another troublesome waste, which was later overcome by this method, was that of synthetic detergents.

Synthetic surfactants became commercially feasible on a large scale in the early 1950s and were readily derived from the raw materials and by-products from the then expanding oil industry. These were broadly grouped into cationic, anionic and non-ionic surfactants. The most widely used was the anionic detergent based upon alkyl sulfonates. This was used by industry and by practically every household in the country. Inevitably, increasing amounts were present in sewage discharges.

The result was that wherever aeration or any agitation occurred during sewage treatment, large banks of foam would be produced. Surfactants retained their foaming properties in concentrations as low as 1.0 mg/L. These surfactants were hardly affected by sewage treatment so that a considerable proportion was discharged into receiving rivers and coastal waters. The widespread use of such materials led to froth and foam being produced on sewage treatment plants with the consequent risk of wind-blown, bacterial infection. Large, unsightly banks of foam were formed on the surface of rivers to which sewage treatment plants discharged. The presence of surfactants in rivers reduced the rate of oxygen absorption at the air/water interface. As little as 0.1 mg/L of detergent will almost halve the uptake of oxygen by a river or stream. Obviously this greatly reduces the self purification capacity of watercourses. The floating banks of foam also reduce sun-light penetration of the water and so retard photosynthesis, causing further deterioration of the watercourse.

The alkyl benzene sulfonates which were almost entirely responsible for this effect had a molecular structure with branched alkyl chains which were resistant to biodegradation. These were known as 'hard detergents'. The molecular formation of a surfactant largely determines its biological degradation potential.

The problems caused by foaming led to the introduction and widespread adoption of new surfactants which compared favourably in cost and cleaning potential to the earlier ones. The replaced surfactants had a 'straight chain' molecular structure and were therefore readily biodegradable. Aerobic biological organisms can now degrade detergents and drastically reduce the damaging effects of foam on rivers (Klein, 1957; p. 6).

TRADE EFFLUENT CONTROL

Increased biological treatment plant capacity had to be made available to treat higher loadings from industrial discharges. This technique was only possible for high organic loads. Persistent toxins, such as metals, were very troublesome. The very bacteria needed to treat the sewage were killed by these toxins and consequently the total sewage flow would not be fully treated. The only practical method of overcoming this was by reducing and pre-treating such discharges at source. More control of trade effluents was needed and this was eventually made possible by legislation.

The Public Health (Drainage of Trade Premises) Act of 1937 gave industries the right to discharge to sewers and a local authority the power to control such discharges. Wastes were allowed to be discharged to public sewers subject to quality, rate of flow and other conditions. The cost of receiving and treating these effluents could also be recovered by the local authority, but all of these controls were only made possible by invoking byelaws. Byelaws were a clumsy instrument with which to exact control and payment for trade effluent treatment; in fact none were passed. Although well-meant, the Act had another disadvantage since it did not apply to any discharge made prior to 1937, granting such discharges a 'prescriptive right'. The majority of discharges existed already and so enjoyed exemption from this legislation.

The outbreak of World War II in 1939 also impeded the enforcement of this Act. Industry needed to concentrate on the war effort and not be restricted by trade effluent controls. Large industries were also employers of local labour and had considerable influence with local councils. Consequently, there was little improvement to trade effluent discharges as a result of the 1937 Act.

It was not until much later in 1961, by means of the Public Health Act 1961 (Part V), that effective control could be exercised over trade effluent discharges to a municipal sewerage system.

This Act not only gave much clearer guidance on conditions which could be imposed upon discharges to foul sewers, but also enabled charges to be made directly by the local authority without resorting to byelaws. There was also provision within the Act to overcome 'prescriptive rights'.

TRADE EFFLUENT CHARGES

This was a turning point in the control of trade wastes and was the first true instance of 'the polluter must pay' principle. Local authorities could afford to establish trade effluent departments to control these wastes at source and many firms had to resort to trade effluent pre-treatment to remove toxins, such as heavy metals, before discharging their wastes to the sewerage system.

Charges made for trade effluent discharges are based on a formula first devised for use at the West Middlesex, Mogden, sewage works. The formula, known as the 'Mogden Formula', compared the strength of the waste, usually in terms of organic load, and suspended solids, with that of domestic sewage. The cost of treating domestic sewage was known and this proportional cost was then applied to the trade effluent, usually in pence per cubic metre.

A portion of the revenue from trade effluent charges was used to finance a controlling inspectorate who would regularly take samples and monitor industrial effluent discharges. The most commonly used instrument of control was the consent document, very few agreements were made with local industry.

Control of the industrial waste discharge at source was really the only practical way to reduce the amount and effect of persistent toxic materials that were discharged to sewage treatment works. The aerobic organisms responsible for sewage treatment were protected in this manner from the damaging discharges of local industry.

The controlling function for these discharges has now passed to the privatised water companies. These companies need to protect their sewage treatment plants from the effects of industrial wastes and so exercise trade effluent control procedures as well as collecting trade effluent charges from industrial users. The consenting procedure is now almost universally applied and the charging formula is still broadly based upon the 'Mogden' principle.

The limits imposed upon discharges to foul sewers within the conditions of a consent document have become more stringent over the years. An example of this would be individual toxic metal limits. When consents were first issued, a limit for an individual heavy metal may have been 50 mg/L. Over the last 10 years, discharges to foul sewers have been limited to <10 mg/L for an individual metal, in some cases <5 mg/L. The current limit for discharges containing cadmium to the foul sewer is generally less than 0.02 mg/L of cadmium.

RIVER POLLUTION LEGISLATION

The polluted state of rivers in England and Wales in the mid-nineteenth century resulted in the appointment of two Royal Commissions on river pollution. The first was appointed in 1865 and the second in 1868. The third report of the 1868 Commission contained such convincing evidence of the deplorable state of the rivers, that legislative action resulted, and the Disraeli government passed the Public Health Act 1875.

The 1875 Act defined the duties of local authorities with regard to disposal and treatment of sewage and established a system of local health administration. The Act stated that sewage discharges should not affect or deteriorate the quality of any water or stream to which it was discharged. Unfortunately sewage treatment techniques had not advanced sufficiently to put these idealistic principles into practice.

A further Act of parliament followed in 1876 which attempted to control the pollution of rivers by sewage and industrial wastes. This was the Rivers Pollution Prevention Act 1876. This Act's relevance to trade effluents is discussed elsewhere and insofar as sewage pollution is concerned the Act made it an offence to put solid matter into a stream (it was necessary, however, to prove that pollution or interference with the flow took place) and the Act also prohibited the discharge of solid or liquid sewage matter into a river. It was no supportable defence that the river was already polluted by sewage from upstream.

This Act was well intentioned, but for all practical purposes was ineffective. The reason for this was that the administration of the law was carried out by the Sanitary Authorities, whose very sewage treatment works were polluting the watercourses. This was changed by the Local Government Act of 1888 which gave similar powers to county councils and later River Authorities, or Boards, were set up in some of the major industrial areas to enforce the Act of 1876. Six River Boards were formed and did valuable pioneering work until the passing of the River's Board Act in 1948 enabled further development of the River Boards.

ROYAL COMMISSIONS

Towards the end of the nineteenth century the improvements in sewage treatment due to the gradual introduction of contact beds, percolating filters and activated sludge plants led to the production of better effluents than those produced from the former land treatment methods. A Royal Commission on sewage disposal was set up in 1898 to report on sewage treatment and trade wastes. This was a major advance towards pollution prevention in this country. This Commission sat for 17 years and produced nine reports.

The fifth report of 1908 reviewed the available methods for treating sewage at that time and concluded that sewage could be purified to any required degree by means of land treatment or artificial filters. The eighth report of 1912 was of great importance, and concerned the quality of sewage effluents which may be discharged to rivers, and to river quality classification. The ninth report in 1915 discusses trade effluent limits for discharge to watercourses and the final report in 1915 summarised the findings of all the previous reports. It is the eighth report of 1912 which is the basis of most consent limits for sewage effluents even today. The standards, based on BOD and suspended solids, were still referred to as 'Royal Commission' standards until quite recent years (Imhoff *et al.*, 1971; p. 310).

THE RIVER BOARDS

The Royal Commission was thus responsible for originating standards for effluents discharged to watercourses and was also instrumental in generating a classification for the watercourses. A central Advisory Water Committee was founded in 1937 to report on the co-ordination of water interests. The reports of this body resulted in the Rivers Board Act of 1948 which established River Boards for the control of land drainage, fisheries and river pollution. These boards were responsible for maintaining or improving the river classifications in their areas and consenting and controlling discharges to watercourses. The majority of these discharges were sewage effluents.

The 1951 Rivers (Prevention of Pollution) Act gave the River Boards the power to consent new discharges with a 'Section 7' consent or by the passing of a Bye-Law. Discharges made prior to 1951 were exempt, a situation similar to the 'prescriptive right' of the trade waste legislation of 1937. In 1960 these powers were extended to tidal waters and the Rivers (Prevention of Pollution) Act 1961 enabled pre-1951 discharges to be controlled and facilitated a more direct consenting procedure (Callely *et al.*, 1977; p. 266).

THE WATER AUTHORITIES

In 1974, as a result of the 1973 Water Act, the Water Authorities were formed. These large, multi-disciplined organisations undertook the duties of the River Boards.

The water authorities were responsible for treating sewage and discharging sewage effluents as well as controlling those effluents and preventing river pollution, which was an unsatisfactory state of affairs. The Water Authorities were both 'poacher' and 'gamekeeper' in this respect, since they policed their own effluents. The effect was similar to that described earlier resulting from the Rivers Pollution Prevention Act of 1876.

This situation prevailed until the formation of the privatised water companies and the National Rivers Authority in 1989. The latter controlled all estuaries and inland waters, as well as discharges made to these, in England and Wales. The National Rivers Authority was subsequently replaced by the Environment Agency in 1996 and all of its powers relating to water pollution control have been transferred to the new body.

RIVER QUALITY CLASSIFICATION

The Royal Commission on Sewage Disposal in their eighth report, volumes 1 and 2, of 1912 and 1913, laid down a classification for river waters and sewage discharges to river waters based upon BOD and suspended solids. Their suggested classification of river waters was 'very clean', 'clean', 'fairly clean', 'doubtful' and 'bad'.

These seemingly arbitrary classifications were based upon general characteristics such as smell, degree of turbidity, presence of fish, presence of suspended matter, algal growth, etc. Such observed phenomena were correlated with averaged analytical data on the water such as BOD testing, oxygen absorbed from potassium permanganate, ammonia, oxidised nitrogen, suspended solids, chloride and dissolved oxygen.

The BOD test was regarded by the Royal Commission as one of the most reliable chemical tests of the quality of river water relative to organic content. Consequently the BOD values established for rivers have been known since as the Royal Commission River Classification. In other words a 'very clean' river would be expected to exhibit a BOD value of 1.0 mg/L, while a 'bad' river would have a BOD value of 10.0 mg/L.

River classification could be expressed in terms of other 'strength' tests once some correlation has been established with the testing parameter over a period of time and under different conditions. The 'oxygen absorbed' value from the potassium permanganate test could equally be used to arrive at a classification, and a river water defined as 'bad' would have a value of 7.0 mg/L of oxygen absorbed (Klein, 1957; p. 13).

The Royal Commission calculated a series of limiting figures for river waters falling between the 'fairly clean' and 'doubtful' classes. A river with such a limiting value would be on the point of showing signs of pollution. The recommendations of the Royal Commission concerning the oxygen balance of polluted waters is summarised as a BOD which must not exceed 4 mg/L and a dissolved oxygen content which should not fall below 4 mg/L (approximately 60% saturation at 65°F, an assumed maximum summer temperature).

Other standards for river classification were developed by River Boards and were very much dependent on local circumstances, such as natural alkalinity or acidity, presence of toxins and the possibility of the presence of miscible and non-miscible solvents or oils.

The limiting figure of 4 mg/L BOD, decided upon by the Royal Commission, provides a basis upon which to fix quality limits for sewage effluent discharges. From knowledge of the river quality upstream or before the sewage discharge, the sewage effluent could be limited in such a way as to produce a mixed river water and sewage effluent quality that does not exceed 4 mg/L BOD. The Royal Commission, in fact, produced formulae to arrive at these levels. These were modified to accommodate dilution factors available in a receiving watercourse. The dry weather flow of the river was known or could be calculated, and by using this flow and the BOD value of river water above the point of discharge, the required BOD of the sewage effluent could be calculated.

Only the principal rivers were investigated by the Royal Commission and these were relatively large and gave more than eight times dilution to received effluent in most cases (i.e. a ratio of one part sewage to eight parts river water). The calculated, desired BOD for such a discharge was 20 mg/L. The Royal

Commission also recommended a standard of 30 mg/L of suspended solids for the same dilution factor, thus most sewage effluents were required to meet a BOD of 20 mg/L and a suspended solids level of 30 mg/L. This became known as the 'twenty-thirty' standard or 'Royal Commission' limit (Klein, 1962). These limits are still imposed today by the Environment Agency on many small sewage effluent discharges. Such was the impact of the Royal Commission's findings that standards recommended in their reports are still in use nearly 100 years later.

Most major discharges have been reviewed by the Environment Agency and tighter limits based on modelled parameters including ammonia and toxic metals are now in force. This method takes into account variable river flows and effluents from other sources in determining limits to protect the use of the watercourse and its potential for recovery at a local level. There are a number of river quality models available for this purpose.

The Environment Agency have introduced Improvements to the River Board classification systems such as General Quality Assessment (GQA) and River Ecosystem (RE) national classification systems which take into account EC Directive requirements and proposed river use. Nutrients are now monitored routinely and biological classification of rivers is also gaining wider acceptance.

CHEMICAL ANALYSIS OF SEWAGE

For the measurement of sewage treatment plant performance, it was essential to arrive at a measurement of sewage strength, and originally various tests were used, all of which depended, like the BOD, on the measurement of oxidisable matter in a specified volume of sample. By determining the amount of oxygen required for the complete oxidation of the organic matter present, an estimation of the strength could be made.

To calculate the oxygen required for complete oxidation of organic matter is difficult and such strength tests are usually based on BOD. In the fifth report of the Royal Commission on Sewage Disposal, McGowan suggested the use of a formula incorporating the potassium permanganate test and ammoniacal nitrogen plus organic nitrogen content, all expressed in parts per hundred thousand. This became known as the 'McGowan Strength'. The McGowan Strength figures can be established for different sewages. For example a weak sewage would have a McGowan Strength of between 60 and 100 parts per hundred thousand, whereas a strong sewage would exhibit a McGowan Strength in excess of 175 parts per hundred thousand.

The McGowan Strength test was used in many sewage treatment works laboratories and was also used for calculating trade effluent strengths in order to levy charges for these effluents. The original Mogden formula incorporated McGowan Strength as a measure of the organic load of a trade effluent and the comparable sewage strength.

The 4-hour N/80 permanganate value tests were also commonly used to establish sewage strength. In this case the oxygen absorbed during the 4-hour incubation period (at 27°C) from acidified potassium permanganate was measured by an iodine titration to give an oxygen absorbed figure in parts per million. The N/80 permanganate values (PV) for a domestic sewage ranged from 50 mg/L for a weak sewage to 150 mg/L for a strong sewage.

Another strength test, which originated in America and became accepted in this country is the chemical oxygen demand (COD). This test measures the oxygen consumed from acidified potassium dichromate over a period of 2 hours.

Unlike the BOD and the COD, the determination of total organic carbon (TOC) is more specific and results can be obtained much quicker than the COD and the 5-day BOD test. The TOC test uses a very small volume of sample which is vaporised. Any organic matter present is oxidised to carbon dioxide which is then accurately measured.

The BOD indicates the amount of organic matter which is biodegradable during the time and conditions of the test, the COD measures the quantity of material oxidisable by the chosen oxidant (usually chromic acid) and the TOC measures the combined value of biodegradable and non-biodegradable organic matter (Klein, 1962).

The methylene blue stability test was a very popular early test used widely to estimate the state, and stability of purification, of sewage effluents. The sample, after the addition of methylene blue, was incubated for a period of 5 days. The absence of a blue colour indicated that the sample was anoxic. Those samples which remained blue were considered to be stable, of good quality and had thus passed the stability test (Isaac, 1957).

The performance of activated sludge plants and bacteria beds could be assessed by measuring the reduction in biochemical oxygen demand before and after treatment. Similarly, assessment could be made of the oxidation of ammonia to nitrites and nitrates.

ANALYTICAL ADVANCES

The imposition of limits to sewage works' effluents by controlling bodies also meant that aerobic sewage treatment processes needed to perform consistently to high standards. The presence of trade effluents in a sewage flow led to the requirement to test for parameters other than strength. Indeed, the presence of toxic trade effluents could inhibit the BOD test producing often meaningless results.

A very wide range of analyses is now applied to sewage and sewage effluents and river waters. Many sewage effluents have limits imposed upon them for not only BOD and suspended solids, but also toxic metals, ammonia, phenols and phosphates. The Environment Agency also investigates river waters and effluents for very low levels of pesticides and chlorinated solvents.

As well as an increased range of analyses being carried out on sewage effluents, the quality of analysis has also improved over the years. Techniques have changed and although in many cases the limits of detection may be no greater, the analysis is more accurate due to improved sample preparation and elimination of interference effects from other substances which may be present.

Additionally, performance testing systems to ensure laboratory procedures conform to universal high standards have been introduced. The United Kingdom Accreditation Service (UKAS) and Aquacheck are two such performance testing bodies. Using sophisticated analytical techniques, levels of less than 0.2 microgrammes per litre (μg/L) can be detected with confidence. For example, the limit imposed on polycyclic aromatic hydrocarbons (PAHs) is currently 0.2 μg/L for water supply purposes.

These advances mean that trace toxins which could adversely affect aerobic sewage treatment plant operation can now be detected. Apparent impaired performance of these units can be attributed to the presence of minute quantities of toxic material. Low level analysis has other implications such as the imposition of more stringent limits upon the quality of discharges to watercourses for example.

QUALITY MONITORING

A more recent adaptation of advances in chemical analysis is that of on-line monitoring of sewage treatment plant and river systems. There are now a wide range of analyses which may be continuously monitored (on-line) and recorded, stored, retrieved or activated remotely.

Sensors are available for a wide variety of determinants based on ion-selective electrodes. Electrodes are available for the measurement of nitrate, sulfide, thiocyanate, ammonia, hydrogen sulfide, chloride, dissolved oxygen, toxic metals and temperature.

The use of continuous measurement of ammonia, total oxidised nitrogen, dissolved oxygen, temperature, organic carbon (UV monitor), suspended solids and pH, enable very detailed, automatic control of biological sewage treatment plants (Pickering, 1977).

Many of these systems are very sophisticated with ultrasonic sensor cleaning and automatic self-calibration at pre-set time intervals to give reference values on a daily or hourly basis. Although expensive, the savings in manpower and the ability to closely and economically control energy levels and effluent quality, make such innovations viable (Kakabadse, 1979).

The ability to confidently measure the strength, or load, that a bacteria bed or activated sludge unit is receiving, enables the calculation of optimum performance to produce the maximum reduction of organic load through the treatment unit. Consequently the advances in analytical chemistry have had direct influences on the design and performance of both bacteria beds and activated sludge plants. Additionally the application of quality limits to sewage works effluents has motivated a need to improve, and in many cases extend, biological sewage treatment within a sewage works, in order to comply with these limits.

Chapter 9

Process development and selection

EFFECTS OF POPULATION INCREASE

Populations of towns increased rapidly over the period 1800 to 1900 mainly due to the Industrial Revolution and the migration of labour to urban areas (Figure 9.1). In absolute terms the urban population tripled between 1850 and 1910 (HMSO, 1957; p. 32).

As the population of towns increased, the land near to towns became more expensive. The dramatic growth of urban areas between 1800 and 1900 has continued to a lesser extent to the present day. This has had inevitable effects on the cost and availability of building land near to towns. Land close to towns became much more expensive than agricultural land.

Such expansion of population proportionally increased the volume of sewage discharged from an urban area and made excessive demands on any existing sewage treatment facilities. Land in close proximity to towns was needed for sewage treatment works construction. Established sewage works needed land to extend and improve sewage treatment to cope with this growing demand. This had the effect of inflating the capital cost of sewage treatment development.

LAND AREA

The land area required for different sewage treatment processes has increasingly influenced process selection. This has been so for a long time and explains the demise of some earlier sewage treatment methods which depended upon large land areas. Treatment processes and their differing requirements for land are shown in Figure 9.2. The land treatment carried out by Franklin and

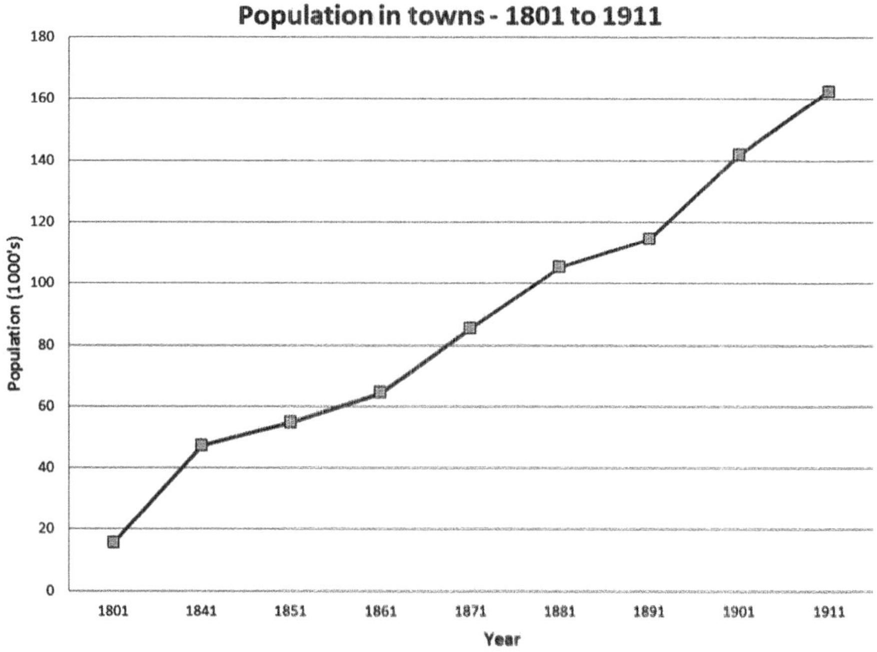

Figure 9.1 Increase in population in towns. Data source: Bedarida (1976); p. 17.

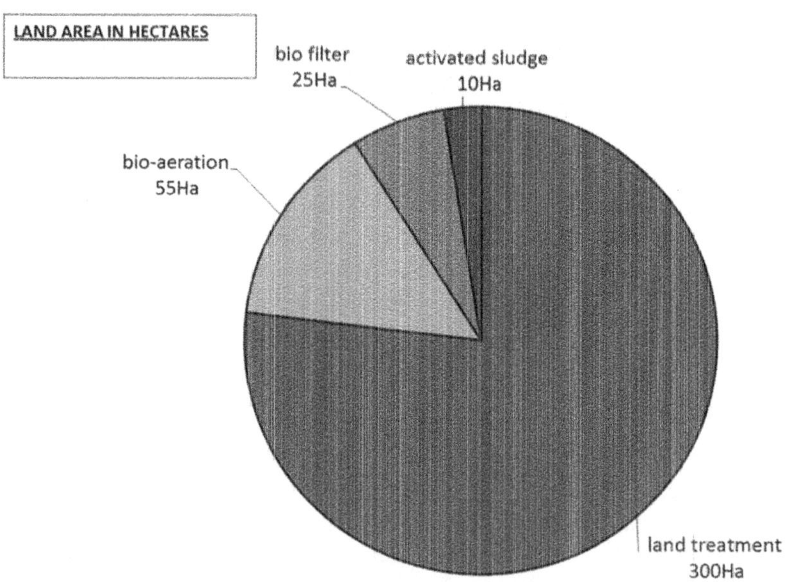

Figure 9.2 Land area requirements to treat sewage from a population of 300 000. Data source: Isacc and Bradler (1969); p. 4.

Bailey-Denton in the 1870s enabled sewage treatment from a town of 300 000 population to be carried out on an area of 300 hectares (Stanbridge, 1976; vol. 6, p. 37).

In fact far more than 300 hectares would have been required in order to 'rest' land areas in rotation. Haworth's Bio-aeration system to treat waste from the same population would occupy 60 hectares. (Institute of Sewage Purification, 1954a, p. 272).

Single filtration through bacteria beds to treat the same flow would take up 30 hectares (Tebbutt, 1983; p. 7). A conventional diffused air activated sludge plant to serve this same population could be built on an area of less than 7.5 hectares (Bedarida, 1976; p. 17).

CAPITAL COSTS

The space occupied by a treatment plant is just one component of a complex equation to select the most suitable process treatment. Capital cost is another important consideration.

There is an obvious proportional increase in cost to treat the waste from larger numbers of people. As a far greater number of smaller treatment plants have been built than larger ones, there is more data available for lower populations. It is also interesting to note that the majority of plants were designed to meet 'Royal Commission' effluent standards (20 mg/L BOD and 30 mg/L suspended solids). Despite the fluctuations in monetary values, such differences in capital cost are still relative.

Capital cost will reflect other criteria to a certain degree and land values will contribute significantly to the overall capital cost of a specific process. Figure 9.3 illustrates the relative capital costs of biological processes (Bolton and Klein, 1971; p. 106). These costs are representative of plants designed to produce a final effluent of 18 mg/L suspended solids and 12 mg/L BOD, and for the nitrifying plant, a level of 6 mg/L ammonia. In this instance (and that of Figure 9.4 below) the biological filter plant produced a nitrifying effluent, meeting the same effluent quality as the nitrifying activated sludge system.

It will be seen that the capital cost of biological treatment is much greater for biological filtration than for activated sludge, (more than double that for non-nitrifying activated sludge treatment), a difference related to necessary land area.

The much lower settlement costs for the bacteria bed system indicate the relatively low production of secondary sludge levels from this process when compared with the more involved activated sludge handling requirement for surplus activated sludge and return sludge facilities.

For the nitrifying plant the longer retention time and higher operational sludge concentration required for nitrification are indicated by the higher costs of the biological stage and the slightly increased cost of the settlement stage when compared to the non-nitrifying activated sludge system.

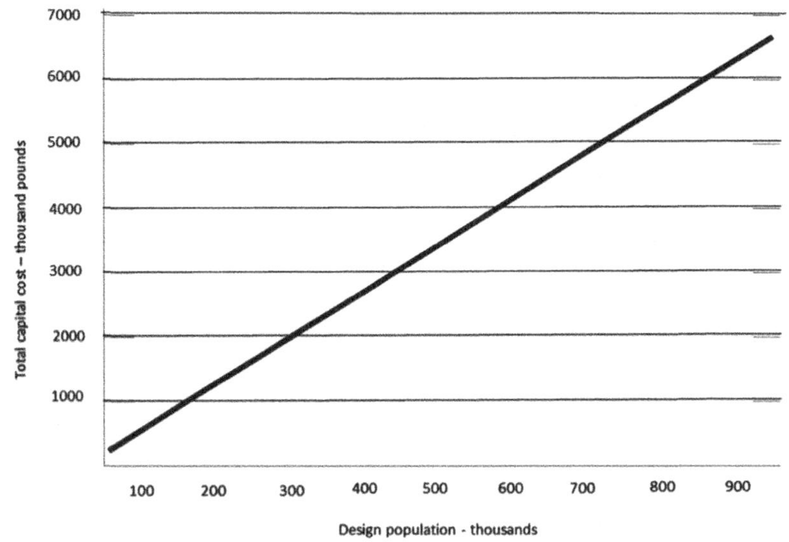

Figure 9.3 Capital costs of biological treatment relative to design population. Data sourced from Spearing (1987).

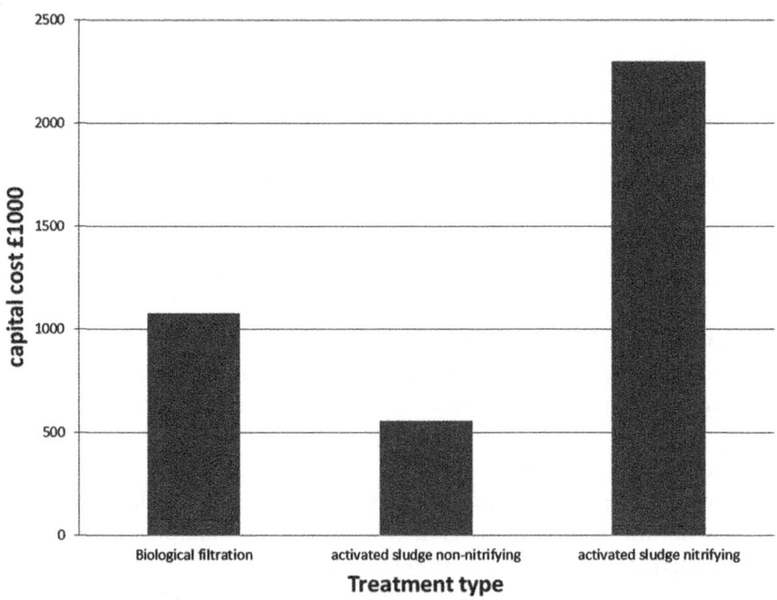

Figure 9.4 Operating costs comparison of biological treatment methods Data sourced from Owen (1982); p. 6.

OPERATING COSTS

Operating costs, including maintenance costs, will also influence process selection, indeed, 'whole life costs' are now often incorporated into plant efficiency calculations to establish the long-term economics of a particular process.

Figure 9.4 compares the operating costs of the same two treatment systems shown in Figure 9.3. The operational costs aptly illustrate that bacteria beds, although expensive to construct, are cheaper to run than activated sludge plants and even have lower secondary settlement and sludge treatment costs. The latter differences are attributable to the difficulties experienced in de-watering and general disposal of activated sludges as well as the higher volumes produced. The principal difference between the two activated sludge systems is almost entirely due to energy costs.

When considering typical budget allocations for energy, labour and adjustments for loan repayment for capital expenditure the costings can be seen to be quite revealing (Isaac and Bradler, 1969; p. 4). This gives an interesting viewpoint as to the relative energy costs of this type of system. Although the expenditure on energy of a smaller plant can be low (less than 10% of the annual costs), larger systems can use as much as one-third of the total annual costs on energy (Figure 9.5). Even the selection of individual components within a treatment

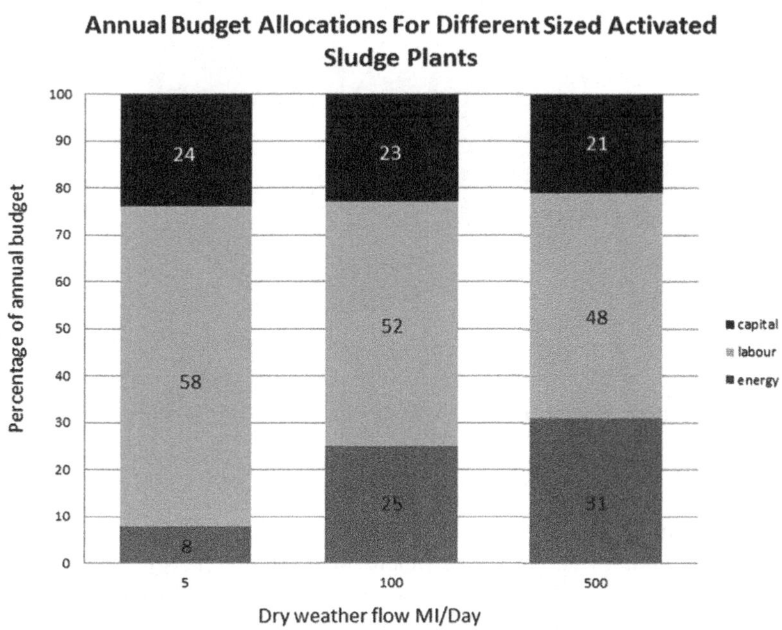

Figure 9.5 Annual cost allocations for different sized activated sludge plants. Data source Spearing (1987).

process assume cost-effective significance when options are available. Consider the method of aeration for a chosen activated sludge treatment. The optimum oxygen transfer rate is desirable but selection is dependent on other factors including energy costs.

An illustration of this is shown in Figure 9.6. Different methods of aeration and their relative oxygenation efficiencies can be compared (Spearing, 1987). The relative efficiency of the rate of oxygen transfer is given in terms of energy requirements. 'Fine bubble' and 'high purity oxygen' are the most efficient and 'coarse bubble' has the lowest efficiency. Both Cryo Pure Oxygen and Pressure Swing Adsorption (PSA) are on-site methods of oxygen production. Essentially these two processes separate the two main components of air (oxygen and nitrogen), and selectively discard the nitrogen, thus providing oxygen or oxygen-enriched air for the purposes of aeration.

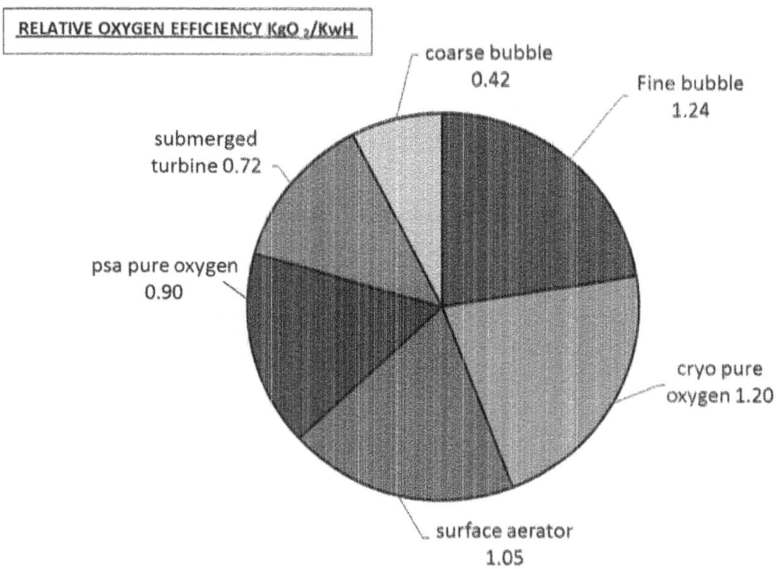

Figure 9.6 Relative oxygen transfer efficiencies. Data derived from Owen (1982); p. 182.

There are other considerations when selecting aeration methods. For example, although 'fine bubble' aeration is efficient in terms of energy, this system has higher capital and maintenance costs than 'coarse bubble' aeration.

Even the configuration of aerators within the aeration tank can significantly effect oxygenation efficiency. A cross-flow mixing pattern has been found to be more effective for 'coarse bubble' aeration than a spiral flow pattern. It has been demonstrated that by optimising tank design, oxygen transfer rates can be increased by up to 50%. (Owen, 1982; p. 6).

RECENT TRENDS

Future treatment systems may use 'designer' bacteria for specific wastes or specific purposes. Specially designed bacteria could be used to accelerate or concentrate sewage treatment or to eliminate undesirable species.

Genetically engineered micro-organisms (GEMs) were first made in the 1960s. Since this time numerous organisms have been constructed for use in different industries (Owen, 1982; p. 182). The drug industry use GEMs to produce hormones, insulin and other agents. The food and brewing industry have used GEMs and they are used in agriculture and forestry. GEM technology is an attractive approach for improving biological treatment. GEMs with desirable and controllable characteristics may revolutionise biological sewage treatment.

Mathematical models and computer aided design will play an increasing role in future developments. A sewage treatment optimisation model (STOM) was developed in the late 1980s which consisted of a set of mathematical models to aid sewage treatment process design relative to cost and plant performance (Bolton and Klein, 1971; p. 106). It is now possible, with programmes such as GPSX produced by Hydromantis Inc., to access a computer simulation which is dynamic (Owen, 1982; p. 183). While a programme is running, one can vary feed flow, loadings, air supply, etc. and even simulate nutrient removal or inhibition. These systems will be more widely used in future to improve the eventual design of a sewage treatment plant and also as an aid to operational management.

System layout, design and application, as well as capital and operating costs, all play an important role in process selection and development. The choice of any specific process for a particular application will consequently depend on many factors and each case will be different. The replacement of one process by another, the abandoning of some older treatment methods and the development of improved techniques is partly explained in terms of performance relative to prevailing economics.

Chapter 10

Conclusions

HISTORICAL OBSERVATIONS

As civilisations become successful and create stable societies with expanding urban developments, the consequent population increase demands an efficient means of sewage disposal.

Historians have tended to confuse sewage disposal with treatment. Early civilisations, such as the Minoans and Romans, provided sewage disposal methods in the form of drainage predominantly for surface water. This should not be regarded as treatment. There was little evidence of sewage treatment even in the eighteenth century in England. Human waste was either thrown onto middens, into the streets or into rivers. In towns the waste accumulated, caused offensive smells and a very real health hazard. In the mid-eighteenthth century such accumulated waste was regularly removed by scavengers for its fertiliser value.

It was the development of early drainage schemes in the major towns to deal with flood water and surface water which led to the adoption of these systems by the local inhabitants to dispose of waste and rubbish. Such drainage relied on rain water to flush obnoxious waste into the nearest rivers. Dry summer periods in towns and cities presented an intolerable health hazard and obnoxious stench.

Human waste disposal has frequently been associated with water. Early public toilets were commonly sited over rivers or streams in order that the waste deposited would be washed away.

The water closet was a logical development as a sanitary means of washing away human waste. Originally for the rich, and discharging to small cess pits or middens, the water closet gained wider use. It was essential for this device to function that a supply of water would be available.

WATER SUPPLY

A water supply is a vital requirement for any urban population. In London, attempts to meet this demand were made by entrepreneurs such as Peter Morris, with his river Thames pumping scheme and Edward Colthurst, who originated the New River scheme.

The provision of a water supply to houses within London and other towns improved dramatically over the period 1670–1820, especially after the advent of steam-powered pumps and iron water mains. With the greater availability of water, the water closet system was more widely adopted.

The supply of water was directed more towards availability and distribution rather than quality and initially little attention was given to pre-treating the water before or after distribution. This could be compared to the transmission and disposal of sewage rather than its treatment.

As more water closets came into use, more human waste was flushed to the nearest watercourse which was usually the source of a potable water supply.

WATER AND DISEASE

The link between the water carriage system for sewage disposal and the waterborne contamination which produced disease in a sewage-polluted water supply, was a crucial point in the development of sewage treatment.

The polluted water supplies resulting from the water carriage system and the primitive water treatments of the time, gave rise to widespread disease. The work of physicians, such as Dr Farr, Dr Boobyer, Dr Snow and Dr Budd conclusively proved the link between disease and water-borne sewage pollution.

HEALTH REFORMS

Great impetus was given to health reforms from which we still benefit today.

Health pioneers such as Sir Edwin Chadwick achieved major advances in sanitation within towns and cities. Ultimately the water and sewerage infrastructure was in place, the water carriage system adopted, and adequate water supply provided. All of these factors led to large volumes of liquid waste being discharged to watercourses near major towns.

It was from this time, when municipal liquid waste was conveyed to one point, that a treatment method, so desperately needed, could be developed.

SEWAGE FARMING

The need to treat this voluminous, almost continuous flow of liquid waste and the recognition that it could be used as a fertiliser, coupled with the profit motive, gave rise to the first development in sewage treatment.

Unfortunately, attempts by James Smith and others to promote sewage irrigation systems as high profit farming methods was not successful. This was due to the high water content of the waste, distribution difficulties, seasonal problems and competition from imported guano.

BACTERIA BEDS

It was not surprising that land treatment led to the development of the bacteria bed. The experiments of Sir Edward Frankland and J. Baily-Denton were successful, but were based upon physical filtration through different media.

Bacteriological advances, particularly the studies made in 1877 by Schloesing and Müntz, and the later work of Warrington, was an important milestone in the development of biological sewage treatment. The efforts of Mills, Crimp, Dibden and Corbett demonstrated a more scientific approach to sewage treatment.

The progress made in the study of micro-biology, experiments undertaken in America and the work of W. Santo Crimp and W. J. Dibdin led to the introduction of the contact bed. Eventually the experimental work of Corbett resulted in the bacteria bed. Key work in this area was undertaken by the Lawrence Experimental Station in Massachusetts and particularly the reports published by Hiram F Mills in 1890.

ACTIVATED SLUDGE

Whereas the progressive development from sewage farming to land treatment and to filtration through soil resulted in the bacteria bed, which was quite logical, the development of activated sludge treatment followed an entirely different route. The combined efforts of Fowler, Mumford, Ardern and Lockett, all working in Manchester, led to this discovery.

The key to activated sludge treatment was the retention and re-use of the suspended solids formed upon the aeration of sewage. The re-aeration of this in admixture with further sewage gave quite remarkable results. The published work of Ardern and Lockett stimulated interest in this mode of treatment and led to many different forms of the activated sludge process.

Awareness of the surface aeration which occurs naturally in rivers led to the evolution of the bio-aeration process by Haworth. His success influenced Bolton who introduced the simplex cone surface aerator. Dr Kestner produced the brush aerator which resulted in the construction of other surface aerators including cage rotors. Brush aeration was exploited in many ways, one of which was the oxidation ditch, a relatively cheap treatment process.

EXCHANGE OF INFORMATION

The value of the many reports of the Royal Commissions on Sewage Treatment must not be underestimated. The Association of Managers of Sewage Disposal

Works, and its successors, made available a vast amount of technical detail and treatment experience. This did much to propagate process treatment techniques throughout the country. The value of such free interchange of ideas and information was incalculable.

Following the privatisation of the water industry and the advent of more powerful commercial forces, there is now less free access to such information. Innovative processes and techniques tend to be jealously guarded.

WASTES FROM INDUSTRY

Many of the problems associated with industrial discharges have now been considerably reduced, predominantly by legislation, control monitoring and treatment at source. Due to the changing nature of industry, the manufacture of new products and use of new materials, vigilance must be continually exercised in this area. The presence in industrial wastes of biologically active products, biodegradable compounds, herbicides and pesticides are just some examples of the need to control industrial effluents.

The Mogden formula has seen changes to its base parameters, but essentially remains the same and is still widely used for trade effluent charge calculations.

CHEMICAL ANALYSIS

The improvements in chemical analysis have been discussed and these can now be very closely allied to process control and water quality monitoring. For monitoring purposes the choice of analysis method or parameter may be imposed by the EU. The 'biodegradability test' would be an example of this. There is an increasing need to use measurements which can be nationally and internationally compared. This is especially so where common EU standards are being sought.

The EU makes much use of the population equivalent, but this is usually related to an oxygen demand test, and can be readily calculated from BOD. It is generally accepted that the BOD test is unsuitable for mechanistic process modelling nor is it easily reproducible. The use of both biodegradable and non-biodegradable COD would possibly have more application.

COMBINATIVE BIOLOGICAL PROCESSES

The activated sludge process and percolating filter have developed into different forms of their original systems and combinations of both systems have been used with success.

Fixed film activated sludge treatment has become a very popular addition to the package treatment plant range. Variants of this process have been introduced successfully on a larger scale for municipal sewage treatment. The aerated, submerged RBC process is an example of this.

It is interesting that development of the two original aerobic processes appears to have culminated in their unification to produce the fixed film activated sludge process.

PROCESS DEVELOPMENT AND SELECTION

No covered bacteria beds have been constructed in recent times for treating waste from large communities, despite their apparent advantages over conventional filters. No Sheffield bio-aeration systems have been installed for many years. These systems have been superseded by more modern, concentrated, effective processes.

As sewage treatment processes have become more sophisticated and process design more flexible, a complex interrelation of factors such as energy costs, land area and capital costs govern process selection for any specific application. There has been a drastic reduction in the land area required for modern systems when compared with their earlier counterparts.

Future trends will include an increased use of mathematical models and dynamic software programmes and possibly more use of 'designer' micro-organisms.

The progress of aerobic sewage treatment over the last hundred years has resulted in a tremendous improvement in plant efficiency and the capability to produce high quality effluent. Inevitably the quest for more compact, more effective and more economical sewage treatment processes will continue.

References

Anon. (1871). *Report of the Committee on the Treatment and Utilisation of Sewage*. Taylor & Francis.

Alan K. (1929). *Sewage Disposal Bulletin*. City of New York.

Barty-King H. (1992). *Water, the Book*. Quiller Press.

Bedarida F. (1976). *A Social History of England 1851–1975*. Methuen.

Bolton J. (1954). The birth of the simplex aeration process at Bury Sewage Works. In: Institute of Sewage Purification, Annual Conference, Blackpool, June, 1954, pp. 401–405.

Bolton R. L. and Klein L. (1971). *Sewage Treatment Basic Principles and Trends* 2nd edition. Butterworths.

Boon A. G. (1976). Technical review of the use of oxygen in the treatment of waste water. *Water Pollution Control Journal*, **Part 2**: 210.

Buer M. C. (1926). *Health, Wealth and Population 1760 to 1815*. G Routledge and Sons Ltd.

Bulloch W. (1936). *The History of Bacteriology*. Oxford University Press.

Callely A. G., Forster C. F. and Stafford D. A. (1977). *Treatment of Industrial Effluents*. Hodder and Stoughton.

Chi-Su Chou C. and Hynek R. J. (1979). Development and performance of air driven. In: RBC's 34th Annual Purdue Industrial Waste Conference, p. 34.

Collins O. C. and Elder M. D. (1977). Experiences in operating the deep shaft activated sludge process. *Water Pollution Control Journal*, 1980, **Part 2**: 272.

Cooper P. F., Hobson J. A. and Jones S. (1989). Sewage treatment by reed bed systems. *Institute of Water and Environmental Management Journal*, **Part 3**: 60–74.

Dekema C. J. and Murray K. A. (1942). Deep enclosed artificially ventilated filter beds v ordinary open filter beds. A comparison under working conditions. *Journal of the Institute of Sewage Purification*, **23**, 311–317.

Doman J. (1929). Results of operation of experimental contact filter with partially submerged rotating plates. *Sewage Work's Journal*, **I**(5): 555–559.

Forster C. F. (1985). *Biotechnology and Waste Water Treatment*, 1st edition. Cambridge University Press, p. 83.

HMSO (1957). *Methods of Chemical Analysis as applied to Sewage and Sewage Effluents*. p. 32.

Hunter A. and Cockburn T. (1944). Operation of an enclosed aerated filter at Dalmarnock Sewage Works. *Journal of the Institute of Sewage Purification*, **Vol 14**: 12–29.

Imhoff W., Muller J. and Thistlethwaite D. K. B. (1971). *Disposal of Sewage and Other Water Borne Wastes*. Butterworth, p. 310.

Institute of Sewage Purification (1930). The Association of Managers of Sewage Disposal Works. *Journal of the Proceedings of the Institute of Sewage Purification*.

Institute of Sewage Purification (1954a). Symposium on the evolution and development of the activated sludge process of sewage purification in Great Britain. *Journal of the Proceedings of the Institute of Sewage Purification*, **Part 3**: 22–26.

Institute of Sewage Purification (1954b). The development of the bio-aeration process. In: Symposium on the Evolution and Development of the Activated Sludge Process of Sewage Purification in Great Britain. *Journal of the Proceedings of the Institute of Sewage Purification*, **Part 3**: 103–114.

Isaac P. C. G. (1957). The treatment of trade waste waters and the prevention of river pollution. In: *Proceedings of a Course held in the Department of Civil Engineering, Kings College, Newcastle upon Tyne*, 1–12 April 1957, p. 14.

Isaac C. G. and Bradler R. M. (1969). *Capital and Operating Costs of Sewage Treatment*. Oriel Press Ltd.

Kakabadse G. (1979). *Chemistry of Effluent Treatment*. Applied Science Publishers Ltd, p. 103.

Kershaw M. A. (1977). A heritage of knowledge and a tradition of service through seventy-five years. *Journal of the Institute of Water Pollution Control*, **76**(2): 138–152.

Kessener H. (1930). Small Bio-Aeration Plants. *Proceedings of the Association of Managers of Sewage Disposal Works*, **11**, 91–94.

Klein L. (1957). *Aspects of River Pollution*. Butterworth.

Klein L. (1962). *River Pollution II-Causes and Effects*. Butterworth, p. 17 Ibid, p. 426.

Lechevalier H. A. and Solotorovsky N. (1974). *Three Centuries of Microbiology*. Dover Publications Inc.

McCallum M. (1968). *Crete-History*. Roumbis S A.

Ousby J. C., Hemming M. L., Plowright D. R. and Walker J. (1977). Deep shaft–latest position. *Water Pollution Control Journal*, **Part 4**: 441.

Owen W. F. (1982). *Energy in Waste Water Treatment*. Prentice Hall Inc.

Parr T. (1989). *1988 Survey of Reed Growth in UK Reed Bed Treatment Systems*. Water Research Centre, Medenham, Kent, Contract report TO.058e1.

Pickering W. F. (1977). *Pollution Evaluation, the Quantitative Aspects*. Marcel Dekker Inc.

Quennell M. and Quennell C. H. B. (1977). *A History of Everyday Things in England.* B T Batisford Ltd. Reynolds R. (1943). *Cleanliness and Godliness*. Allen & Unwin, p. 100.

Royal Commission on Metropolitan Sewage Discharge. (1984). *First Report of Commissioners*.

Spearing B. W. (1987). Sewage treatment optimisation model – STOM – the sewage works in a personal computer. *Proceedings of the Institution of Civil Engineers*, **Part 1**: 1161.

Sprague de Camp L. (1963). *The Ancient Engineers* Souvenir Press, p. 164.

Stanbridge H. H. (1976). *History of Sewage Treatment in Britain.* Institute of Water Pollution Control.

Tebbutt T. H. Y. (1983). *Principles of Water Quality Control.* Pergamon Press.

Tilley D. F. (1970). Coventry Civic Affairs, No 241. Coventry Corporation, p. 4.

Wishart J. M. and Wilkinson R. (1941). Purification of settled sewage on percolating filters in series with periodic change in the order of the filters. *Journal of the Institute of Sewage Purification*, **69**, 39–58.

Wylie J. C. (1959) *The Wastes of Civilisation.* Faber & Faber.

IWA Publishing's authorised EU representative for General Product Safety Regulations is Diane D'Arras, 15 rue Duret, 75116 Paris, France, e-mail: safety@iwap.co.uk.

Printed and bound by CPI Group (UK) Ltd, Croydon, CR0 4YY
12/05/2026
02108870-0009